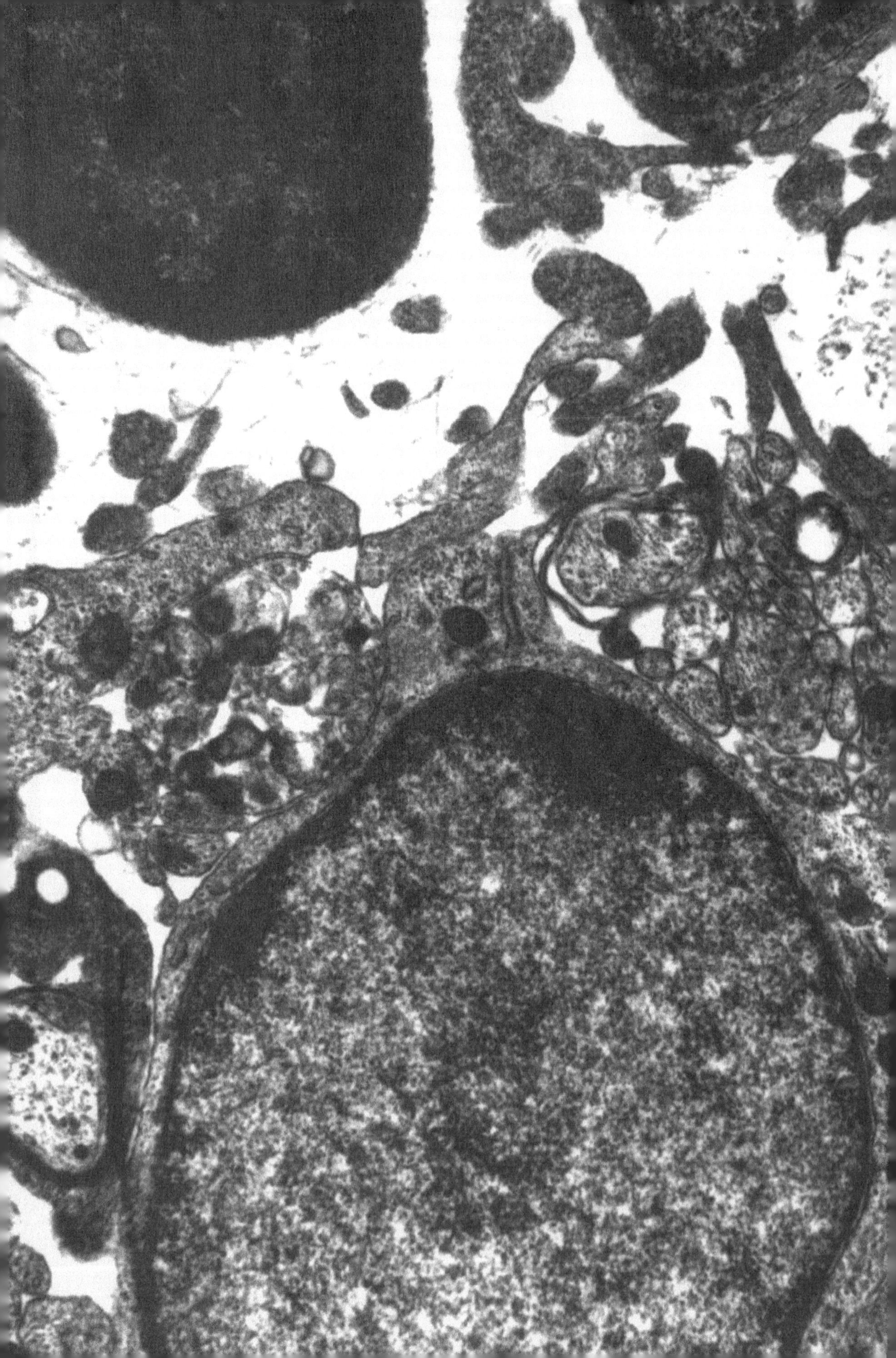

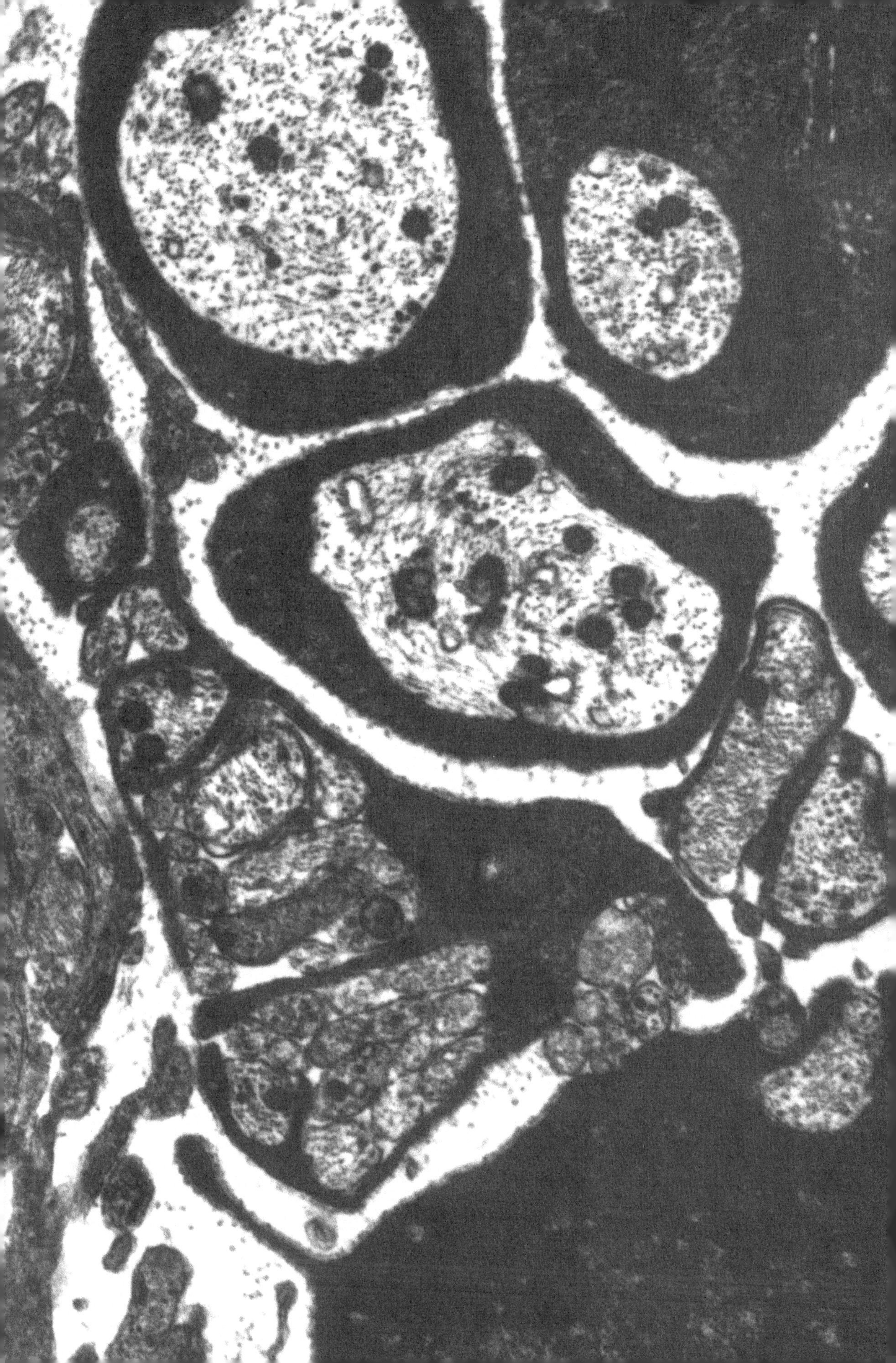

Neuroscience
Intelligence
Unit

# Cholecystokinin and Anxiety: from Neuron to Behavior

Jacques Bradwejn, M.D., F.R.C.P.C.
University of Toronto
Toronto, Ontario, Canada

Eero Vasar, M.D., Ph.D.
University of Tartu
Tartu, Estonia

Springer-Verlag Berlin Heidelberg GmbH

R.G. Landes Company
Austin

# NEUROSCIENCE INTELLIGENCE UNIT

## CHOLECYSTOKININ AND ANXIETY: FROM NEURON TO BEHAVIOR

R.G. LANDES COMPANY
Austin, Texas, U.S.A.

Submitted: June 1995
Published: August 1995



U.S. and Canada ISBN 978-3-662-21707-8


Originally published by Springer-Verlag, Heidelberg Germany 1995
Softcover reprint of the hardcover 1st edition 1995



International ISBN 978-3-662-21707-8



**Library of Congress Cataloging-in-Publication Data**

Cholecystokinin and anxiety: from neuron to behavior / [edited by]
Jacques Bradwejn, Eero Vasar.
p. cm. — (Neuroscience intelligence unit)
Includes bibliographical references and index.
ISBN 978-3-662-21707-8 ISBN 978-3-662-21705-4 (eBook)
DOI 10.1007/978-3-662-21705-4
1. Anxiety—Pathophysiology. 2. Cholecystokinin—Psychotropic effect.
I. Bradwejn, Jacques, 1952– . II. Vasae, Eero. III. Series.
[DNLM: 1. Anxiety—physiopathology. 2. Choelecystokinin–physiology. WM 172
C547 1995]
RC531.C46 1995
616.85'22307—dc20
DNLM/DLC 95-548
for Library of Congress CIP

# Publisher's Note

R.G. Landes Company publishes five book series: *Medical Intelligence Unit, Molecular Biology Intelligence Unit, Neuroscience Intelligence Unit, Tissue Engineering Intelligence Unit* and *Biotechnology Intelligence Unit.* The authors of our books are acknowledged leaders in their fields and the topics are unique. Almost without exception, no other similar books exist on these topics.

Our goal is to publish books in important and rapidly changing areas of medicine for sophisticated researchers and clinicians. To achieve this goal, we have accelerated our publishing program to conform to the fast pace in which information grows in biomedical science. Most of our books are published within 90 to 120 days of receipt of the manuscript. We would like to thank our readers for their continuing interest and welcome any comments or suggestions they may have for future books.

Deborah Muir Molsberry
Publications Director
R.G. Landes Company

# CONTENTS

# EDITORS

**Jacques Bradwejn, M.D., F.R.C.P.C.**
Professor of Psychiatry
Head, Psychobiology and Clinical Trials Research Unit in Anxiety
The Clarke Institute of Psychiatry
University of Toronto
Toronto, Ontario, Canada
*Chapters 4, 9*

**Eero Vasar, M.D., Ph.D.**
Professor and Head,
Insitute of Physiology
University of Tartu
Tartu, Estonia
*Chapters 3, 9*

# CONTRIBUTORS

Nick J. Coupland, D.M., M.R.C.P., M.R.C. Psych
Psychopharmacology Unit
School of Medical Sciences
Bristol, Avon, United Kingdom
*Chapter 1*

Jacqueline N. Crawley, Ph.D.
Chief, Section on Behavioral Neuropharmacology, Experimental Therapeutics Branch
National Institutes of Health
Bethesda, Maryland, U.S.A.
*Chapter 6*

Valérie Daugé, Ph.D.
Université René Descartes (Paris V)
Département de Pharmacochimie Moléculaire et Structurale
Paris, France
*Chapter 8*

Jaanus Harro, M.D., Ph.D.
Department of Pharmacology
University of Tartu
Tartu, Estonia
*Chapter 3*

Diana Koszycki, Ph.D.
Psychobiology and Clinical Trials Research Unit in Anxiety
The Clarke Institute of Psychiatry
Toronto, Ontario, Canada
*Chapter 5*

# CONTRIBUTORS


Pekka T. Männistö, M.D., Ph.D.
Department of Medical Pharmacology
University of Uppsala
Biomedical Centre
Uppsala, Sweden
*Chapter 3*

Finn Cilius Nielsen, M.D., D.M.Sc.
Department of Clinical Biochemistry
Rigshospitalet
Copenhagen, Denmark
*Chapter 2*

David J. Nutt, D.M., M.R.C.P., M.R.C.Psych.
Psychopharmacology Unit
School of Medical Sciences
Bristol, Avon, United Kingdom
*Chapter 1*

Lars Oreland, M.D., Ph.D
Department of Medical Pharmacology
University of Uppsala
Biomedical Centre
Uppsala, Sweden
*Chapter 3*

Jens F. Rehfeld, M.D., D.M.Sc. & Sc.
Department of Clinical Biochemistry
Rigshospitalet
Copenhagen, Denmark
*Chapter 2*

Bernard P. Roques, Ph.D.
Professor, Université René Descartes (Paris V)
Director, Département de Pharmacochimie Moléculaire et Structurale
Paris, France
*Chapter 8*

Franco J. Vaccarino, Ph.D.
Professor of Psychology and Psychiatry
University of Toronto
Mood and Anxiety Division
The Clarke Institute of Psychiatry
Toronto, Ontario, Canada
*Chapter 7*

# PREFACE

It has recently been postulated that cholecystokinin (CCK) might play a role in the neurobiology of anxiety and/or panic attacks. With this book we intend to inform basic and clinical scientists of animal and human research supporting this hypothesis. In chapter 1, N.J. Copland and D.J. Nutt introduce this subject by addressing general issues of neurobiological research on anxiety and panic and in chapter 2, J.F. Rehfeld and F.C. Nielsen review the general properties of CCK as a neuropeptide.

The next three chapters review animal research on CCK and anxiety (J. Harro, E. Vasar, L. Oreland and P. T. Männistö), neurobiological investigations with CCK-4 in humans (J. Bradwejn) and the role of psychological factors in response to CCK-4 in humans (D. Koszycki). In chapter 6, J.N. Crawley addresses the subject of interaction of CCK and other neurotransmitter systems in the understanding of the role of this peptide in anxiety. In two complementary chapters, F.J. Vaccarino discusses the relationships between CCK, addiction and anxiety (chapter 7) and V. Daugé and B. Roques, the neurobiological and behavioral contrasts of the opioids and CCK systems (chapter 8).

In the conclusion, we summarize the main findings and major questions on the role of CCK and anxiety and discuss future direction in research.

*Jacques Bradwejn and Eero Vasar*

# CHAPTER 1

# NEUROBIOLOGY OF ANXIETY AND PANIC

**Nick J. Coupland and David J. Nutt**

## WHAT SHOULD NEUROBIOLOGICAL THEORIES OF ANXIETY AND PANIC EXPLAIN?

Anxiety disorders can be subdivided into groups according to various criteria, the most reliably recognized is the presence of panic attacks, which are made distinctive by their sudden onset and intense somatic symptoms. Phobias are characterized by the relationship of anxiety to cues and avoidance behavior. Post-traumatic stress disorder is distinguished particularly by intrusive thoughts, images or dreams related to specific traumatic experiences. In addition to being able to explain these defining features, a comprehensive neurobiological theory should also make sense of associated features, the natural history of the disorders, putative etiological factors such as life stresses, family history, genetic loading, substance abuse, the association of anxiety with depression, the effects of treatments and findings in physiological, pharmacological and psychological tests. To date, our understanding comes mainly from findings in cross-sectional comparisons between diagnostic groups, with some longitudinal data from treatment studies, and we have to rely largely on animal data to fit this into a theoretical framework. For example, the group of Gorman et al has proposed a neuroanatomical model for panic disorder, with processes in the brain stem, limbic system and prefrontal cortex

*Cholecystokinin and Anxiety: from Neuron to Behavior,* edited by Jacques Bradwejn and Eero Vasar. 

being related to panic attacks, anticipatory anxiety and phobic avoidance respectively.[1] The main evidence in man for separating these has come from treatment studies, which show that antipanic effects can be dissociated from effects on anticipatory anxiety or avoidance. Some work has shown the importance of a history of panic attacks across diagnostic categories in determining panic responses to sodium lactate,[2] and has compared patients with panic with or without agoraphobia,[3] but there has been little study designed to dissect out which features of the disorders are underlined by which abnormality. The advent of functional and receptor neuroimaging techniques should aid progress by allowing the examination of regional rather than global changes.

## ROLE OF CHALLENGE TESTS

Given the limited possibilities so far for directly examining the brain function in man, the challenge test has been utilized as a way of testing hypotheses about central function in psychiatric disorders. Panic disorder has been particularly intensively studied, because of the early realization that panic attacks could be provoked by an exogenous substance[4] and the later demonstration that a variety of challenges may produce anxiogenic effects.[5] Although the provocation of panic has been a fruitful area of research, which constitutes the bulk of this chapter, there are difficulties with this approach. Firstly, there is evidence that cognitive factors complicate the interpretation of behavioral effects in this type of paradigm (discussed below). Secondly, the relationship between the biological stimulus and its physiological consequences may be affected by the behavioral responses. For example, if a test provokes anxiety it can be uncertain whether physiological responses are due to altered sensitivity to the challenge agent, or are a secondary consequence of the anxiety. For this reason, supporting evidence from studies which do not involve anxiety provocation are important in establishing etiological significance for positive findings. There are also some difficulties with the criteria which have been proposed as validating challenge-induced responses as a model of panic[6-7] (see chapter 4). Although such criteria are helpful in promoting intellectual rigor, there are important practical constraints on their application. Firstly, it is required that the putative panicogenic agent should not just trigger the somatic symptoms of a panic attack, but also the affective state of anxiety. Although this is superficially a simple requirement, the degree of anxiety which is accepted as a criterion for rating panic has varied widely between studies, from minimal[8] to intense.[9] We would argue that a truly valid panicogen should be *capable* of provoking severe anxiety in patients, as intense fear or discomfort is the defining symptom for panic disorder. However, in terms of the practicalities of human research, this presents immediate difficulties for validating the use of drugs with a relatively prolonged duration of action, such as yohimbine or fenfluramine, as the provocation of prolonged severe anxiety is not ethical. Given that panic disorder patients do have mild as well as severe attacks,[10-11] it seems reasonable to accept a trade-off between the potency and duration of action of the agent and the stringency of the criterion. A threshold of moderate anxiety may be the only practical criterion for some studies. However, we do not think that having slight or minimal anxiety as a criterion will ever tell us more about panic than to try a higher dose in future studies.

Secondly, it has been stated that the agent should provoke attacks that resemble the patients' usual attacks. This is potentially misleading when comparing the effects of an exogenous substance with a natural attack and we should not jump to the conclusion that a particular transmitter system should be excluded etiologically on the basis that the resemblance is not complete. For example, yohimbine-induced panic attacks differ from spontaneous attacks in that yohimbine may produce lacrimation, rhinorrhea and euphoria.[12] However, oral administration of an $\alpha_2$-adrenoceptor antagonist will lead to slowly increasing effects on noradrenergic neurons in multiple central and peripheral regions, which could lead to many irrelevant actions in addition to a relevant one. If undue weight is given to the irrelevant actions there is a risk of disregarding the important finding. Given that natural panic is of rapid onset, intravenous bolus dosing is more likely to model it than oral dosing, but we cannot inject drugs into a specific brain region of panic patients to reproduce physiological events perfectly.

We would agree that panicogenic effects should be relatively specific for patients with a history of panic attacks and that the effects should be reproducible, but a final caveat may be expressed in relation to the criteria that effective antipanic treatments should, but ineffective antipanic agents should not block the effects of the agent. This presupposes that treatments work by correcting some etiological factor, rather than compensating for it, and that a panic model is identical to a natural panic attack. In our view, the mechanisms of both have to be considered. For example, imipramine acutely may provoke panic and chronically is an effective antipanic agent,[13] and yohimbine may provoke panic. Since both drugs increase synaptic noradrenaline, this supports the hypothesis that an abnormality of noradrenergic function is involved in panic. However, giving an $\alpha_2$-adrenoceptor antagonist during chronic reuptake inhibitor treatment blocks two of the main controls over synaptic noradrenaline. Preclinical microdialysis studies suggest that after chronic desipramine treatment the percentage increase in noradrenaline release is decreased in frontal cortex, suggesting a decreased presynaptic $\alpha_2$-adrenoceptor effect; in fact the absolute levels of noradrenaline released are higher, because reuptake is also blocked.[14] This could indeed lead to worse anxiety rather than attenuation.[15] Data from studies where tricyclics are augmented by yohimbine tend to support this possibility, as many depressed patients became more anxious and agitated on the combination.[16-17]

Some additional tests may also be helpful in validating a model. For example, where a drug has both central and peripheral actions, the panicogenic effect should be blocked by antagonists which have access to relevant sites in the brain, but not by those which do not. Finally, evidence that the agent has a role in anxiety in other species would be helpful, although the uncertain validity of animal models of panic means that again this cannot be dogmatic.

## A ROLE FOR ABNORMAL RESPIRATORY PHYSIOLOGY?

The initial observation by Pitts and McClure[4] that the intravenous infusion of sodium lactate provokes panic attacks in patients who suffer from panic, together with the incorporation of panic disorder as a syndrome in the DSM classifications has led to extensive subsequent research on the mechanism of action of sodium lactate. Lactate-induced panic attacks resemble spontaneous attacks in symptom profile, although patients and controls who do not panic also have substantial somatic symptoms. Responses which would be atypical for naturally occurring panic, such as a surge in cortisol, are not generally produced. The panic attacks can be reliably reproduced in those who do panic, although a substantial minority of patients do not, and they are attenuated by a range of effective treatments (see Table 1.2).[18] They have good specificity for patients with panic disorder compared with healthy controls.[2] In comparisons with patients with other psychiatric diagnoses, the main factor influencing specificity appears to be the history of panic attacks: patients with other primary diagnoses have increased panic responses to lactate if they have had panic attacks, but not if they have not (see Table 1.1).[2] A number of theories have been developed to try to explain the mechanism by which lactate produces its effects, which have been reviewed in more detail elsewhere.[5,18] Some of the suggested mechanisms can be refuted in the light of existing data: secondary hypoxia is unlikely as this has not been found to be an anxiogenic stimulus; hypocalcaemia can be induced by other agents without panic; metabolic alkalosis is unlikely both because bicarbonate is a less potent panicogen than lactate, despite producing a similar degree of alkalosis and because alkalosis alone would be expected to lead to a compensatory hypoventilation, $pCO_2$ increase and respiratory acidosis, but in fact patients and controls hyperventilate and reduce their $pCO_2$.[5] It is possible that patients are hypersensitive to the $CO_2$ generated and this provokes the attacks, which is supported by the effects of breathing $CO_2$ itself, as described below. Other central effects of lactate are also possible, particularly as the D isomer of lactate can also provoke attacks, despite it not being metabolized to $CO_2$.[19]

The provocation of panic attacks in panic disorder by $CO_2$ itself has been validated.[20] It induces psychological and somatic

***Table 1.1 Specificity of response to different panicogenic agents across the anxiety disorders***

| test | controls | panic disorder | PTSD | social phobia | GAD | simple phobia |
|---|---|---|---|---|---|---|
| lactate | + | ++ | (+++) | + | + | ? |
| $CO_2$* | + | ++ | ? | + | + | + |
| yohimbine* | + | ++ | ++ | ? | + | ? |
| mCPP* | + | ++ | ++ | ? | + | ? |
| flumazenil | 0 | +++ | 0 | +? | +? | ++? |
| CCK* | + | +++ | ? | ++ | (+?) | ? |
| caffeine* | + | ++ | ? | ? | ? | ? |

?: not known; 0: no panic attacks; + minority panic; ++ about half panic; +++ majority panic; *: known to be dose dependent (indicates positive effects, but patients also had panic attacks); PTSD: post-traumatic stress disorder; GAD: generalized anxiety disorder symptoms, which are attenuated by antipanic treatments. It shows reasonable specificity for panic disorder patients over controls and other anxiety disorders, although this is a threshold phenomenon and there is a trade off between the reliability of panic provocation, which may require higher doses, and specificity, as increasingly high doses will provoke panic in increasing proportions of healthy subjects.[3,20] The sensation of dyspnea alone is unlikely to explain the findings as dyspnea induced by other methods, such as increasing airway resistance, does not provoke panic as potently.[21]

The possibility that increased sensitivity to $CO_2$ links these two types of panic provocation has led to the hypothesis that these are manifestations of an overly responsive suffocation alarm, which may act spontaneously or when triggered by inappropriate cues.[22] The dyspnea, smothering or choking sensations, hyperventilation and urge to escape, which are defining symptoms of panic attacks, are then understandable physiological and psychological responses to the threat of suffocation. Certainly hyperventilation appears more appropriately placed in the context of being a symptom than as having a central role in causing panic. Firstly, hyperventilation is not a necessary concomitant of naturally occurring attacks, as demonstrated by ambulatory $CO_2$ monitoring,[23-24] although a subgroup of patients may be chronic hyperventilators.[18] In addition, the extent to which voluntary hyperventilation reproduces sensations associated with panic appears largely to be determined by suggestion[25-26] and to have little relationship to whether or not patients hyperventilate during naturally occurring panic.[23]

The false suffocation alarm hypothesis suggests a relatively specific response to certain kinds of respiratory stimuli. Given that patients clearly respond to a wider range of internal and external stimuli,[1,5] this might be explained by conditioning, either to associated environmental cues or to associated somatic symptoms, such as a pounding heart. An alternative viewpoint is that the abnormality lies in a more general alarm system, which might respond initially to a wide range of triggers.[5]

## NORADRENERGIC SYSTEM

The noradrenergic system has been suggested as serving such a function.[27] The evidence for this is partly based on animal studies, which demonstrate activation of the locus ceruleus and increased noradrenaline turnover and release during stress, anxiety-like responses during increased activity of the locus ceruleus, sensitization of the locus ceruleus to repeated stress and attenuation or blockade of these effects by anxiolytic treatments.[27] Recent evidence suggests that locus ceruleus activation is involved physiologically in mediating the level of attention which is given to stimuli, according to characteristics such as aversiveness, novelty or salience.[28] When the locus ceruleus is artificially stimulated, at modest levels this appears to enhance the focus of attention, but high activity leads to labile, scanning attention.[29]

Adrenergic and noradrenergic function may be examined in humans by measuring the peripheral or cerebrospinal fluid (CSF) concentrations of the catecholamines or their metabolites, or through challenge tests. Although consistent changes in catecholamine levels have not been demonstrated during panic attacks, at least some are associated with elevated methoxyhydroxy phenylglycol (MHPG), the main metabolite of noradrenaline.[30] Increases in MHPG are also seen in social phobics and volunteers exposed to public speaking stress.[31-32] These findings indicate increased activity of the sympathetic system, but could represent a feature of nonspecific stress or of anticipatory anxiety rather than of panic. Plasma noradrenaline or adrenaline have not been consistently altered during panic attacks, for example those induced by lactate. We do not see sympathetic activation as being an invariable feature of panic however, as the tachycardia and palpitations could be due to vagal withdrawal.[33] In addition, negative results from peripheral measures do not in themselves exclude a role of central noradrenaline. The locus ceruleus, which is the source of noradrenergic innervation of the forebrain, does not project directly to preganglionic sympathetic neurons. Plasma noradrenaline and most MHPG is produced peripherally and inferences about central function from plasma levels are made on the assumption that the locus ceruleus and sympathetic system change in parallel,[34] when they may be dissociated under many circumstances, such as postural change. The fact that in exposure studies subjects are upright and physically active may mask any component of sympathetic activity which is related to central arousal. There is however, evidence for sustained elevation of peripheral catecholamines in PTSD.[35-36]

When directly manipulating noradrenergic activity in panic disorder through challenge testing, using β-agonists, such as isoprenaline, or the $\alpha_2$-adrenoceptor agonist, clonidine or antagonist, yohimbine, abnormalities have emerged. Isoprenaline precipitates panic in some patients.[37] However, as it does not cross the blood-brain barrier and the patients have downregulated peripheral β-receptors,[38] the affective component of the panic may be mediated via conditioning or cognitions. The $\alpha_2$-adrenoceptor antagonist, yohimbine, has been shown to induce panic attacks in up to two thirds of panic disorder patients or to produce greater increases in anxiety than in controls in a number of studies, although the occurrence of panic is not particularly robust with the oral route.[39-41] When yohimbine is given intravenously, anxiety peaks within 5 minutes of the end of the infusion and patients who do panic show an increased MHPG response compared with controls.[39] Panic patients also have a greater sympathetic activation on standing after yohimbine than controls.[41] Several studies have shown blunted growth hormone responses and in some cases increased MHPG and blood pressure responses and decreased sedative responses to clonidine in panic disorder.[42] This supports the belief that the yohimbine-induced effects are not just secondary to panic, but may indicate an underlying abnormality in the noradrenergic system. The strongest evidence to date has come from a study in which patients were challenged with both yohimbine and clonidine on separate occasions. Those patients who panicked after yohimbine were the ones who also had the smallest growth hormone responses to clonidine.[39] It might be supposed that increased noradrenaline release is leading to down-regulated, post-synaptic responses to clonidine. However, blunting of responses to growth hormone releasing hormone (GHRH) have also been reported in panic disorder,[43] so that it should be tested whether or not susceptibility to yohimbine-induced panic also predicts blunted GHRH responses, before concluding that the clonidine response is mediated by abnormal, post-synaptic $\alpha_2$-adrenoceptor function. It may be that another factor such as corticotropin releasing factor is altering responses to both stimuli.[44]

Increased responses to yohimbine are not entirely specific to panic disorder, as it also provokes panic and symptoms such as flashbacks in some PTSD patients, even in the absence of a history of panic attacks,[45] but patients with GAD or obsessive compulsive disorder (OCD) do not panic.[46-47] A blunted growth hormone response to clonidine has also been found in GAD and social phobia, without significant differences in cardiovascular or biochemical effects from controls.[42] Together with the lack of difference between GAD subjects and controls in responses to yohimbine, this suggests that although there could be increased activity in noradrenergic projections to the hypothalamus in these disorders, this is not due to an abnormality in presynaptic control.

Yohimbine-induced anxiety in panic disorder is blocked by antipanic treatment with alprazolam[48] and fluvoxamine, although interestingly the latter does not appear to attenuate the MHPG and cardiovascular responses.[49] Increasing 5HT function inhibits the response of the locus ceruleus to one of its main excitatory inputs, the glutamate projection from the nucleus prepositus hypoglossi, which might explain how the central effect is dissociated from the peripheral sympathetic response.[50] The difficulties of using imipramine treatment to validate the yohimbine model have been described above. One new measure which has shown promise in overcoming some of the complexities in interpreting noradrenergic responses, is the use of $\alpha_2$-agonists during sleep. Clonidine potently suppresses REM sleep, an effect which is dose-dependent, clearly central (probably presynaptic) and due to the fact that the patient is asleep and is unlikely to be mediated cognitively. Administration of clonidine during sleep has provided an elegant demonstration that $\alpha_2$-adrenoceptor sensitivity is reduced in major depression and also that there was no evidence for a similar abnormality in a GAD control group.[51]

Noradrenergic disregulation is unlikely to be an all-encompassing explanation for the symptoms of panic disorder. Firstly, it has been pointed out that yohimbine-induced panic leads to cortisol release, whereas spontaneous panic does not.[23] This argument is not conclusive, however, as a similar cortisol release occurs in healthy controls.[39] This could mean that although spontaneous- and yohimbine-induced panics both involve increased locus ceruleus firing, yohimbine has the additional effect of removing the inhibitory $\alpha_2$-adrenergic control on cortisol release.[52] Secondly, clonidine reduces locus ceruleus firing, but although clonidine pretreatment attenuates lactate-induced panic, it does not fully block it.[53] This argument can however be reversed, on the basis that lactate may act via sensory stimulation, driving the locus ceruleus through its glutamate input.[50] In normals, negative feedback of noradrenaline on presynaptic receptors will limit the extent to which locus firing increases, whereas if patients have reduced presynaptic $\alpha_2$-adrenoceptors, neither endogenous noradrenaline or an exogenous agonist will exert such an effect. When clonidine was tested as an antipanic treatment, panic attacks broke through again after a short period, supporting the idea that receptor down-regulation may limit its effect.[54] Other situations in which the locus is overactive in man, such as opiate withdrawal, have a different clinical picture to panic attacks, although again this may be because there is tonic rather phasic activation, plus additional changes in other systems.[50] Perhaps the most important argument is that some patients do not have abnormal responses either to yohimbine or to clonidine.[39] Noradrenergic disregulation may be implicated in symptoms such as heightened arousal, excessive attention to threat and difficulty focusing attention elsewhere. That sudden increases in central noradrenergic activity form the sole mechanism of panic is less likely.

## BENZODIAZEPINE RECEPTOR SYSTEM

The benzodiazepine site is an allosteric modulatory site on the GABA-A receptor and it is now well known that agonists potentiate GABA-mediated chloride influx and hyperpolarization, leading to in vivo effects such as anxiolysis and sedation, whereas inverse agonists decrease GABA actions and are anxiogenic.[55] Evidence has accumulated that GABA modulates the output of many of the systems thought to be involved in anxiety,[56] leading to a number of possible hypotheses as to how alterations in function might be

***Table 1.2. Effects of different treatments on the panicogenic effects of challenge tests***

| test | TCA | SSRI | MAOI | BDZ | CBT | non treatment |
|---|---|---|---|---|---|---|
| lactate | ++ | ? | ++ | ++ | ++ | 0 |
| $CO_2$ | ++ | ? | ? | ++ | ++ | ? |
| yohimbine | 0 | ++ | ? | ++ | ? | ? |
| mCPP | ? | ? | ? | ? | ? | ? |
| flumazenil | ++? | ++? | ? | 0? | ? | ? |
| CCK | ++? | ++ | ? | ++? | ? | ? |
| caffeine | ? | ? | ? | ? | ? | ? |

?: insufficient or absent data; 0: not blocked; + slight; ++ moderate; +++ full blockade;
TCA: tricyclic antidepressant; SSRI: selective serotonin reuptake blocker;
MAOI: monoamine oxidase inhibitor; BDZ: benzodiazepine; CBT: cognitive behavior therapy

involved in anxiety states. Anxious patients might produce more of an anxiogenic, endogenous inverse agonist or less of an agonist. Alternatively, GABA-benzodiazepine function could be altered, such that the activity of ligands is shifted towards the inverse agonist position, as occurs in tolerance.[55]

There is some evidence for the production of endogenous inverse agonists in man, with the example of tribulin, which can be extracted from human brain and urine, and which inhibits benzodiazepine binding. Increased tribulin excretion has been associated with post-traumatic stress disorder[57] and generalized anxiety disorders.[58] Even though an endogenous neurotransmitter has not been definitively identified, the availability of benzodiazepine agonists and antagonists allows some hypotheses about their possible actions to be tested in humans. It has been demonstrated that exogenous inverse agonists do provoke anxiety,[59] although ethical considerations have prevented sufficient testing to decide whether this models panic well. Although the possibility of prolonged severe symptoms could be overcome with flumazenil, the proconvulsive and kindling properties of inverse agonists do not recommend them for human research.[60]

Two studies, using an antagonist and an agonist, have provided the strongest evidence that the benzodiazepine site may be involved in panic disorder. Our group found that flumazenil 2 mg iv provoked panic attacks in eight out of ten panic disorder patients, but in none of the controls.[61] All panickers rated their anxiety as severe or very severe, the onset was rapid and in general the effects were rated as similar to natural attacks, except that respiratory symptoms were less prominent. The findings could not be attributed to concurrent or recent alcohol or benzodiazepine abuse. In a study including a small number of panic disorder patients, flumazenil-induced panic did appear to involve respiratory symptoms and further work is needed to clarify whether or not this represents a systematic difference from spontaneous attacks.[62] In the second study saccadic eye movements were used to quantify sensitivity to the agonist, diazepam.[63] It is well established that benzodiazepines cause a decrease in peak velocity, acceleration and deceleration, together with an increase in errors, when saccadic eye movements are made to a moving target. These effects are reversed by flumazenil.[64] Panic disorder patients have normal baseline saccade parameters. However, they were subsensitive to the effects cumulative doses of intravenous diazepam on saccades, with the average concentration of diazepam which induced a 30% slowing in peak velocity being three times higher in the patients than in controls.[63] In other studies, panic disorder patients have shown lower sensitivity to diazepam in terms of its effects on plasma noradrenaline kinetics and amnesic effects,[65] but the saccade paradigm is particularly striking as the responses are not affected by baseline differences or subject motivation. Further validation for the measure being related to benzodiazepine sensitivity has been provided by showing that chronic alprazolam treatment is associated with further blunting.[66]

Other research which should be mentioned includes an inconclusive flumazenil study. Several patients did have panic attacks, but unreliable bioavailability and slow onset of effects via the oral route may have lowered the panic rate.[67] The study in which flumazenil 1 mg iv was given to anxiety patients treated with benzodiazepines illustrates a further problem with using the effects of treatment to validate a panicogen. Patients became anxious after flumazenil and two with a history of panic disorder had panic attacks, but if there is a shift in the benzodiazepine spectrum, then clearly treatment with benzodiazepine agonists may be compensating for it rather than reversing it. A challenge which then leads to acute benzodiazepine withdrawal on top is unlikely to be blocked.[62] We have preliminary findings in a cohort of patients treated effectively with antidepressants that none of the four on treatment had a panic attack, but one of the two who had stopped did panic. Overall the rate of panic was therefore 17% and the mean change in panic inventory scores was lower than in patients with active panic disorder (Fig. 1.1).

The clearest conclusion that can be drawn from these studies is that panic is unlikely to be due to an endogenous inverse agonist, as flumazenil would be expected to block this. Given that flumazenil does not attenuate the effects of CCK-4 in volunteers,[68] it is also unlikely that CCK-induced panic is mediated by an inverse agonist. With the evidence for subsensitivity to agonists, the results in panic disorder suggest that there may be a shift in the spectrum of benzodiazepine site function towards an inverse agonist position. Given that a number of functions, including memory and noradrenergic function appear to be affected, this may be across a number of brain regions. Alternatively, an endogenous agonist could be produced in panic disorder in an attempt to counteract anxiety and, although it does not completely suppress anxiety, the symptoms become far worse when displaced by the antagonist. This may seem unlikely, but there is a parallel in pain studies, where sufferers with rheumatic pain have shown evidence in a recent PET (positron emission tomography) study using $^{11}C$ diprenorphine, that increased endogenous opiate binding due to thalamic receptors may occur, without achieving full analgesia.[69] To date, although studies have suggested possible reduced iomazenil binding to the temporal cortex in panic disorder, these results have to be interpreted cautiously, due to methodological problems such as partial volume effects.[70]

We have been examining the specificity of the behavioral findings with flumazenil, by using the same paradigm in other anxiety

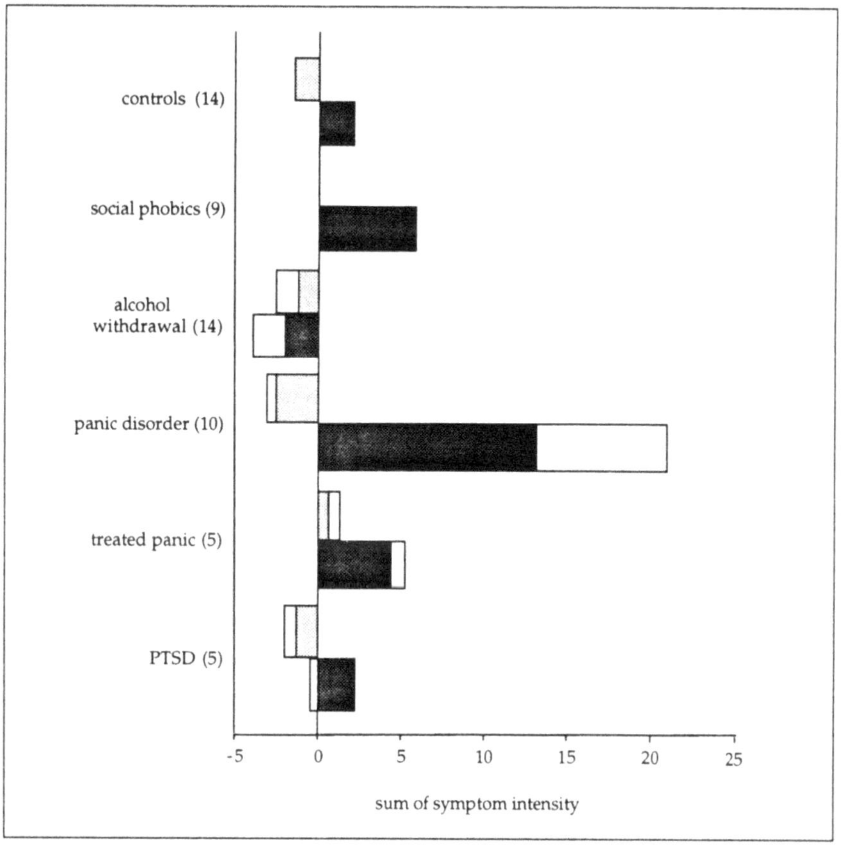


*Fig. 1.1. Specificity of flumazenil. Median panic symptoms after intravenous flumazenil 2mg (dark shading) and placebo (light shading). Scores are of the number of symptoms multiplied by their peak intensity and are separated into somatic symptoms (shaded areas) and anxiety and fears (unshaded).*

disorders (see Fig.1.1). We have used a criterion for rating panic of an increase in score of ≥ 2 in 4 or more symptoms from a panic symptom inventory, together with anxiety self-rated as severe or very severe, and showing a rapid increase from baseline. Symptoms are rated from 0 (not at all present) to 4 (very severe), administered at baseline and then as a rating of peak effect made 5 minutes after flumazenil or when the symptoms have settled, if this takes longer.

Thus far, we have not found that flumazenil provokes panic in post-traumatic stress disorder and in fact it may ameliorate anxiety, which is similar to the findings of Charney et al (personal communication). This clearly argues against flumazenil-induced panic being due to baseline state anxiety, which was higher in our PTSD group than the panic subjects. Likewise only 1 of 14 subjects with acute alcohol withdrawal panicked after flumazenil, despite marked baseline anxiety. In social phobics there have been more varied responses, with 2 out of 9 subjects tested having had panic attacks, which were shorter than naturally occurring panic. It should be remembered that social phobics may never suffer spontaneous or cued panic attacks, as some subjects suffer from shame, embarrassment or blushing, together with anxiety, rather than with panic. Larger numbers will therefore be needed to see if responses vary with previous experience of panic attacks.

## SEROTONERGIC SYSTEM

A number of challenge procedures have been used to assess the 5HT systems, although the answers provided have been rapidly outstripped by new questions raised by the explosion in receptor subtypes which have been identified. Attempts to alter 5HT availability with l-tryptophan, 5-hydroxytryptophan, or with tryptophan depletion do not provoke anxiety or panic in untreated patients.[71-73] mCPP, which has mixed agonist properties at some subtypes and antagonism at others, and the 5HT releasing agent, fenfluramine, have both provoked panic attacks in challenge paradigms, although some authors have described the patients' responses to fenfluramine as being more akin to anticipatory anxiety than true panic.[74-75] The response to mCPP is also not clear-cut. Some studies have not found the anxiogenic responses to be specific for patients compared with controls.[76] It may be that this is a threshold phenomenon, with lower doses providing a clearer separation,[77] or an additive effect, with more patients having enough symptoms to meet panic criteria because of their higher baseline anxiety, rather than to increased sensitivity to the drug. The $5HT_{1A}$ partial agonist, ipsapirone, provoked slight feelings of nervousness in panic disorder patients, without frank panic attacks.[78]

Why should stimulating 5HT receptors by some methods lead to anxiety and others not? At least some of these discrepancies may be due to L-tryptophan and D-fenfluramine releasing 5HT differentially from dorsal and median raphe projections. Both increase plasma prolactin, but L-tryptophan responses are blocked by the $5HT_{1A}$ antagonist, pindolol, whereas the D-fenfluramine responses are blocked by $5HT_{2C}$ receptor antagonists, which appear in animals to be preferentially associated with dorsal raphe projections.[79] The dorsal raphe innervates the amygdala and periaqueductal grey. Deakin has proposed a hypothesis that the amygdala projection is involved in conditioning and in inhibiting approach towards aversive stimuli, akin to anticipatory anxiety, whereas the projection to the periaqueductal grey inhibits panic until the threat becomes immediate. In addition to animal evidence, there is some human volunteer research to support this, in that the $5HT_2$ antagonist, ritanserin, reduces conditioned responses to an aversive stimulus of loud noise, whereas it increases fear and autonomic responses to the unconditioned stress of unprepared public speaking, which in this context is suggested to be a model of panic.[80] Antidepressant treatment is theorized to block panic attacks by increasing the inhibitory serotonergic transmission to the periaqueductal grey, but to be of little direct benefit on anticipatory anxiety because it increases the serotonin drive to the amygdala. Theoretically, ritanserin should improve the anticipatory anxiety, by blocking $5HT_2$ transmission to the amygdala, but as it could also release the periaqueductal grey, this might explain why in practice patients do not benefit.[81]

## OTHER PANICOGENS

These will not be discussed in detail, but findings are summarized in Table 1.1 and 1.2, in other reviews,[5,12] and covered in detail for CCK and pentagastrin in chapter 4.

## COGNITIVE CONSIDERATIONS

The idea that pharmacologically provoking panic attacks can shed light on neurobiological mechanisms in panic disorders has been criticized by cognitive theorists, who have proposed a variety of psychological explanations of panic disorder.[82-83] It is important that the limitations of these explanations should be stated: although some theorists see the cognitive features of anxiety as part of multifaceted disorders, the view that they are the whole explanation can be used to deny the value of research into the neurobiology of anxiety. The core of cognitive theories of panic is well known, that the disorder is caused by patients evaluating the threat posed by bodily symptoms as being much more severe than it really is: "... panic attacks *result* from the catastrophic misinterpretation of certain bodily sensations. The sensations which are misinterpreted are mainly those which are involved in normal anxiety responses (e.g., palpitations, breathlessness, dizziness etc.) but also include some other bodily sensations." A sequence of events is postulated, with triggering stimuli (either internal sensations or external cues) leading to a perceived threat, which causes apprehension and anxiety-related bodily sensations. These are then catastrophically misinterpreted, leading to a spiral of apprehension and anxiety, culminating in a panic. Panic will not develop in the absence of catastrophic misinterpretations.[83] Different routes have been proposed as to how panic disorder patients develop these faulty evaluations. Some authors have proposed that they are predisposed, with a personality trait, named "anxiety sensitivity," of holding a more negative set of beliefs about bodily symptoms.[84] An alternative view is that patients develop a panic response as a result of repeated associations between sensations, catastrophic thoughts and intense anxiety, leading to the response becoming increasingly automatic, a process called "interoceptive conditioning."[85]

There is little dispute that patients may have fears of disastrous consequences during panic attacks, or that such worries may heighten their anxiety or become a stressor which helps to perpetuate the disorder. On the other hand, several lines of evidence dispute a central etiological role of catastrophic cognitions in panic disorder. Firstly, not all panic attacks involve catastrophic cognitions and in students with unexpected attacks, the panic symptom profile is indistinguishable between those who report them and those who do not.[86-87] As pointed out by Costello, studies may also inflate the frequency with which catastrophic misinterpretations occur, by not clearly distinguishing them from sensations.[88] For example, in the study of Street et al, high correlations were reported for the occurrence of the "sensations" of "feeling faint... choking... and nausea" with the "cognitions" "will pass out... will choke... will throw up."[89] In another survey patients endorsed their most intense symptom as "a sensation of doom, as if something terrible will happen."[90] When patients are responding to this type of questionnaire, are they really stating that they had both a sensation and a cognition, or are they endorsing alternative descriptors for the same event? To assess cognitions further, panic disorder patients were given questionnaires in which they had to give an explanation of why they might have various sensations. Fourteen to thirty-eight percent of their interpretations were of the catastrophic kind (compared to 4-21% for controls), whereas for 66-78% of vignettes their interpretations related to them feeling anxious.[91-92] In a further study, subjects were given a forced choice between suggested interpretations, which were carefully selected to avoid referring to anxiety reactions.[93] If only the patient-control differences for catastrophic misinterpretations are examined in this way, they would appear to support the theory, as patients are significantly more likely to endorse them. However, if patients are allowed to choose that they may be anxious, the data suggest that they are much more likely to attribute their sensations to anxiety than they are to make a catastrophic misinterpretation. Even when panic disorder patients have thoughts about harmful consequences, they may still recognize that these are due to anxiety. Although high ratings were found for patients' belief in their catastrophic cognitions during attacks (averaging 84 on a 0-100 scale for strength of belief) in one study,[94] we found a mean of 31 in 15 patients assessed at our clinic.[95] This may reflect differences in the patient populations, which could be influenced by referral bias, the more cognitively apparent panickers being sent to psychologists. Studies in unselected populations such as in general practice would be useful in clarifying this.

A critical observation against the catastrophic misinterpretation theory is that panic disorder patients can also awake from particular stages of sleep in a panic attack, suggesting a relationship with a physiological process rather than cognitions or responses to arousal.[96] Thus, although catastrophic thoughts do occur and treatment studies suggest a maintaining role in panic disorder, in our view they cannot play the pivotal role that has been suggested. This assessment has been queried on the basis that catastrophic misinterpretations are not always evident because they can be unconscious. This makes the theory difficult to refute, but there are several observations which are inconsistent even with an unconscious process. Although panic attacks wax and wane, they tend to persist despite the lack of a catastrophic outcome. Seligman has pointed out that this contravenes what would be expected under learning or conditioning theories.[97] In order to account for this difficulty, it has been suggested that panic patients never really test whether the catastrophic outcome will occur, so that they never learn.[98] For example, they may engage in overt behaviors (holding on to their shopping trolley to prevent fainting) or special thoughts to prevent disaster. However systematic evidence has not been provided to support this and when asked directly how they get through their attacks, patients were as likely to state that the anxiety just got better by itself as to give any other reason.[99] The effects of medication for panic disorder are also difficult to explain on the basis of unconscious cognitions. The β-blocker, propranolol, reduces anxiety-induced tachycardia and tremor, symptoms which could be misinterpreted, but is ineffective.[100] It has been argued that central symptoms, such as dizziness, depersonalization, or arousal, which would not be blocked by propranolol, can be misinterpreted as well as peripheral ones.[82] The effects of imipramine and specific serotonin reuptake inhibitors are difficult to incorporate into a theory where misinterpretations of central arousal are critical. These drugs may increase arousal initially and do not decrease anticipatory anxiety as effectively as they do panic.[13] Patients say they think that an attack is just about to commence, but nothing further happens. It is not clear from cognitive theory why this arousal is no longer catastrophically misinterpreted. It could be argued that medication is providing a safety cue, but this would not explain the ineffectiveness of propranolol, or why effective treatments show dose-response relationships.[101] Successful noncognitive treatments, such as imipramine or reflective listening, are associated with reductions in scores on ratings of catastrophic cognitions.[102] In the case of imipramine, this improvement was lost after a period off treatment during which a number of patients relapsed.[103] This pattern strongly suggests either that cognitions change as a result of clinical state, or that the measures are not very good at distinguishing cognitions from other symptoms.

Recent studies using computerized tests of cognitive biases have also found drug treatment to normalize abnormalities in panic patients, supporting the first explanation.[95]

In our view the suggested role for cognitive misinterpretation in panic attacks during challenge tests, that panic will occur only in the presence of catastrophic cognitions, is also unsustainable.[83] Provocation of catastrophic cognitions during challenge tests is a necessary component of the cognitive argument, but not sufficient proof in itself, as they could again be the result rather than the cause of the panic attack. Nevertheless, if subjects experienced panic attacks without reporting catastrophic cognitions, this would be difficult to accommodate in a cognitive explanation. Koszycki et al have reported the effects of CCK-4 in healthy volunteers, who were stratified into groups with low, medium or high scores on the Anxiety Sensitivity Index. The ASI score did not predict the frequency of panic attacks in the groups, although it did predict the frequency of catastrophic cognitions.[104] The study has been criticized on the basis of a reanalysis using the additional criterion that subjects had to have a catastrophic cognition to define a panic attack.[105] With this definition, the high anxiety sensitivity group had significantly more panic attacks. The circularity of this argument is obvious (the tendency to worry about anxiety symptoms predicts worry about drug-induced anxiety symptoms). A requirement for catastrophic cognitions to be an obligatory part of a definition of panic attacks would ensure that cognitions play a central role in panic. Some authors have used such a definition, whilst accepting "minimal" anxiety.[8] Since sudden onset intense anxiety with four other symptoms are the defining characteristics of panic attacks, in our view it is safer to use this criterion when trying to understand their etiology, rather than restricting attention to a subgroup of attacks involving cognitions.

A further suggested test of the cognitive hypothesis was that cognitive manipulations will "block" challenge-induced panics.[83] Studies with healthy subjects have shown that manipulating their expectations or interpretations of their responses can affect their reactions. This is however variable: there are both positive and negative reports of the effect of instructional set on responses to voluntary hyperventilation[106-107] and effects on reactions to 35% $CO_2$ were weak although statistically significant.[107] In a contrast between subjects instructed to expect unpleasant feelings or pleasant excitement, three of six healthy volunteers in the latter group still reported anxiety after sodium lactate and the statistical difference rested on one subject with a large positive response.[108]

In studies in panic disorder, Rapee, Mattick and Murrell, gave patients either minimal or detailed instructions about what symptoms to expect from a 35% $CO_2$ inhalation. Panic attack rates were not given, but the detailed explanation group rated the feeling of panic as less intense, their response overall as less similar to their usual panic and reported fewer catastrophic cognitions.[109] It is possible that the lower similarity to their usual panic may be partly a response bias, from being instructed to attribute symptoms to the gas, rather than wholly due to a difference in the experience, but assuming that both groups received a similar dose, the evidence supports their responses to $CO_2$ being influenced by instruction. In what has become a landmark study, when prolonged inhalation of 5% $CO_2$ was used, one group given the false belief that they could switch it off had a 20% panic rate and the group which thought it had no control had an 80% panic rate.[8] The definition of a panic attack required that a catastrophic cognition must be present, rather than having a threshold for severity of anxiety. Peak anxiety did not differ significantly between the groups, although it was numerically higher in the "no control" group. The most common cognition reported in the "no control" group was "fear of losing control." The startling difference between the panic attack rates has often been quoted, but may have been influenced by the definition used. A further study of 5.5% $CO_2$ did not find that attempting to manipulate safety cues altered panic rates.[110] The debate has also been fueled by a number of unpublished studies of panic disorder, in which groups have claimed that controlling the instructional set did not influence panic attack rates in response to prolonged 7% $CO_2$ inhalation,[22] but that it did alter the rates in response to hyperventilation or to sodium lactate infusion.[111] On the available data we would say that manipulations influence responses, but that the degree to which they do so has perhaps been exaggerated.

This takes us to a further point, that findings from challenge studies have shown differing potency between panicogens in the same patients, such as $CO_2$ and CCK-4,[112] and dose-response relationships for individual panicogens.[113-114] This could affect the degree to which cognitive predisposition or manipulation influences the results. It may be that this is only the case for weaker panicogenic stimuli. For example, although anxiety sensitivity was a weak predictor of responses to $CO_2$[3] or hyperventilation-induced symptoms,[115] it had no strong or significant correlations with anxiety induced by CCK-4.[116] There is a potential dilemma here, in that by aiming to produce only a modest biological activation, either to avoid maximal distress for ethical reasons, or show a threshold difference between patients and volunteers, the effect may well be more susceptible to cognitive manipulation. This may lead to any biological component being discounted inappropriately. The method used by Donnell and McNally may be one way of checking the role of expectancy without performing manipulations. They included response bias items in their assessment of symptoms induced by hyperventilation and used these as covariates for analysis of panic inventory items.[115]

Some challenge test results cannot readily be explained on a cognitive basis. For instance, although controls were no better than chance at discriminating yohimbine 20 mg orally from placebo, patients showed significant increases in anxious and panicky feelings, who discriminated the drug.[40] How could panic patients misinterpret stimuli which cannot be discriminated? A suggested cognitive explanation is that panic disorder patients may have greater "interoceptive awareness," that is an ability to detect internal sensations.[111] Although some experimenters have found panic patients to be slightly better than controls at estimating their heart rate,[117] the majority of studies have been negative[118-120] and others have also shown panic patients to have poor or variable ability to judge their airway resistance.[117] In fact the best evidence of an increased ability to detect stimuli has come from a study in which panic patients were more aroused, had more awakenings and had larger heart rate responses than healthy controls to low doses of lactate administered during deep sleep.[121-122] The term "awareness" implies a conscious event, whereas this response suggests an alteration in a preconscious process.

The cognitive approach also does not readily explain why differences between panicogens occur and why some challenge agents have not provoked panic, if the panic symptoms are due to the patient's interpretation rather than a biological effect. For example, it was noted that flumazenil-provoked panic was associated with fewer respiratory symptoms that the patients' usual attacks,[61] and it

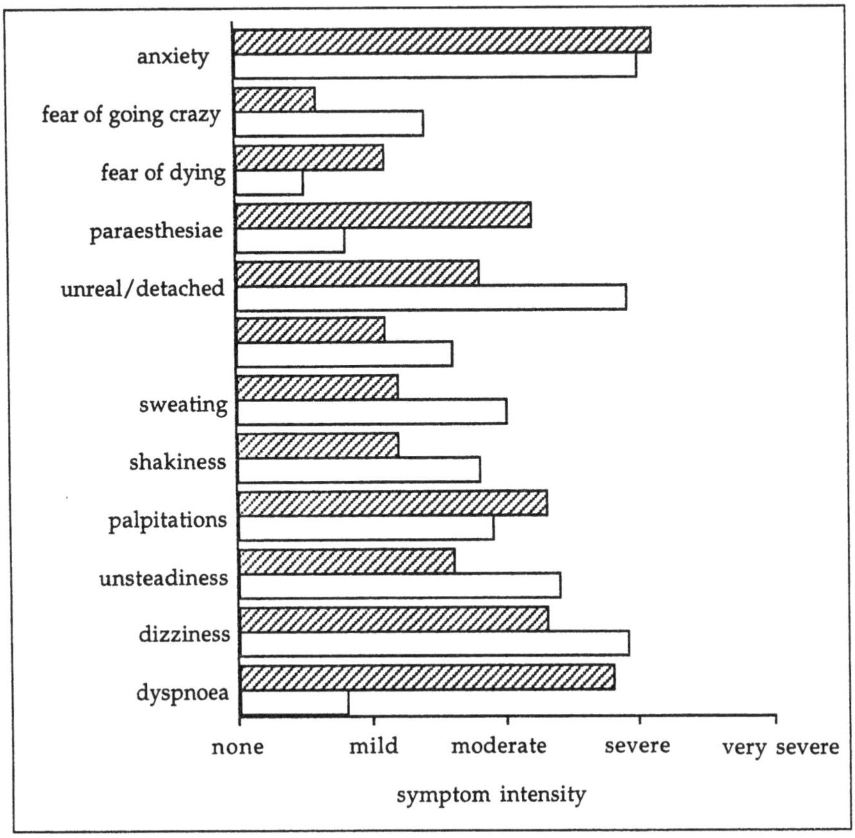


*Fig. 1.2. Profiles of CCK-4 and flumazenil induced symptoms in panic disorder patients. Comparison of mean scores on individual panic items reported in two separate groups of patients. CCK-4: hatched bars (From Bradwejn J et al. Biol Psychiatry 1992; 32:903-912) and flumazenil: unshaded bars (From Nutt DJ et al. Arch Gen Psychiatry 1990; 47:917-925).*

appears less likely to produce dyspnea and paresthesia than CCK (see Fig. 1.2). If the patients are simply panicking in response to a noxious event, it might be expected that they would have their usual symptoms. Similarly other challenges provoking unpleasant symptoms and arousal, such as insulin-induced hypoglycemia, EDTA or physostigmine should also provoke panic but they did not.[12] Thyrotropin releasing hormone (TRH) produces some symptoms, such as a palpitations, nausea and paresthesia, which are similar to those of the panicogens CCK and pentagastrin (see chapter 4),[123] yet only 5% of panic disorder patients given TRH 500 µg had a panic attack.[125] One possibility is that differences in the instructions to patients create different expectancies.[111] However, there are other possibilities, which we have started to explore in healthy volunteers by administering TRH (500 μg) and pentagastrin (0.3 or 0.6 μg/kg) as bolus doses over 5 seconds. These doses of pentagastrin have induced panic in 50-70% of panic patients in other studies, with insufficient data to show a dose-response.[123-124] The main difference so far is that pentagastrin induces chest discomfort, dyspnea and more intense paresthesia, even at the lowest dose, whereas TRH leads to more urinary urgency. The overall intensity of symptoms was similar between TRH and the lower dose. Given the short duration of the cardiovascular responses, we recorded heart rate and blood pressure using a Finapres monitor, which is a non-invasive, beat-to-beat recorder. This measure shows a greater tachycardia and peak pressor effect for the higher dose of pentagastrin, but quite similar effects to TRH at the lower dose (see Fig. 1.3). If this pattern is confirmed in a full sample, it would suggest the hypotheses that pentagastrin produces a higher panic rate than TRH either because of a specific biological sensitivity,

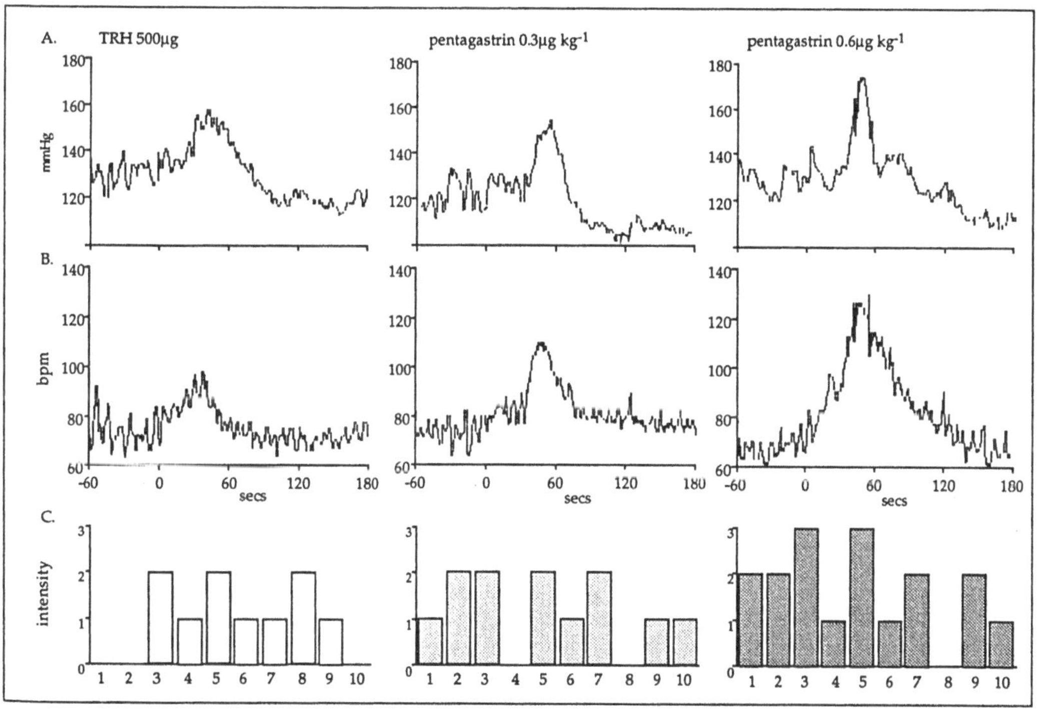


*Fig. 1. 3. Comparisons of beat-to-beat (a) systolic BP and (b) heart rate, together with (c) panic inventory scores after TRH 500 μg or two doses of pentagastrin in a healthy volunteer given at time 0 seconds.*
*Key to panic symptoms: 1. Short of breath. 2. Discomfort in chest. 3. Pounding heart. 4. Sweating. 5. Nausea. 6. Feelings of unreality. 7. Tingling. 8. Urinary urgency. 9. Apprehension. 10. Anxiety.*

or because of specific symptoms such as dyspnea, rather than its cardiovascular effects or the general intensity of symptoms. Tests in patients using a dose of TRH as an active control, matched for cardiovascular effects or overall symptom intensity with pentagastrin or CCK, would therefore be a further way to test these hypotheses with cognitive factors controlled.

## THE FUTURE

Having concentrated on psychopharmacological approaches in humans, we do not wish to leave it unsaid that further advances will also come from other areas, such as molecular genetics, or from our increasing understanding of responses to stress, given the probable role of both of these etiological factors in anxiety.[126-127] It is clear that although psychopharmacological research has demonstrated a number of biological changes which occur during anxiety disorders, there remains a great deal to be done in order to fit this into an overall theory. The available evidence suggests that a number of different neuronal systems may be involved and that no single finding, whether biological or cognitive, can explain all the features of a disorder. Developments in neuroimaging offer some hope that in future we will be able to get a better grasp on how different regions of the brain may be co-ordinated in producing anxiety responses. This will include activation studies that have already been used to examine provoked symptoms in lactate-induced panic, simple phobias and obsessive-compulsive disorder. With the greater possibility of multiple studies using functional MRI it may become possible to control better the nonspecific stress of the imaging procedure by exposure and to extract common features of responses to multiple panicogens. At a more subtle level it may become possible to examine differences in brain activity which may underlie the cognitive biases and more general slowing of information processing in panic disorder patients.[96] In addition radioligand studies will allow us to examine some of the regional neurotransmitter controls over anxiety and in collaboration with the British Medical Research Council's cyclotron unit, our group's initiatives in this direction have begun with the development of techniques for quantifying benzodiazepine binding and also for measuring receptor occupancy by an agonist and its functional effects together, within a single scanning session.

ACKNOWLEDGMENTS

Our thanks to Dr. Andrea Malizia and to Dr. Aviv Weinstein for their helpful discussion of this paper.

REFERENCES

1. Gorman JM, Liebowitz MR, Fyer AJ et al. A neuroanatomical hypothesis for panic disorder. Am J Psychiatry 1989; 146:148-161.
2. Cowley DS, Arana GW. The diagnostic utility of lactate sensitivity in panic disorder. Arch Gen Psychiatry 1990; 47:277-284.
3. Rapee R, Brown TA, Anthony MA et al. Response to hyperventilation and inhalation of 5.5% carbon dioxide-enriched air across the DSM-III-R anxiety disorders. J Abnorm Psychol 1992; 101:538-552.
4. Pitts FM, McClure JN. Lactate metabolism in anxiety neurosis. N Engl J Med 1967; 277:1329-1336.
5. Nutt D, Lawton C. Panic attacks: a neurochemical overview of models and mechanisms. Br J Psychiatry 1992; 160:165-178.
6. Gorman JM, Fyer MR, Liebowitz MR et al. Pharmacologic provocation of panic attacks. In: Meltzer HY, ed. Psychopharmacology: a third generation of progress. New York: Raven Press, 1987:980-983.
7. Guttmacher LB, Murphy DL, Insel TR. Pharmacologic models of anxiety. Compr Psychiatry 1983; 24:312-326.
8. Sanderson WC, Rapee RM, Barlow DH. The influence of an illusion of control on panic attacks induced via inhalation of 5.5% carbon dioxide-enriched air. Arch Gen Psychiatry 1989; 46:157-162.
9. Liebowitz MR, Fyer AJ, Gorman JM et al. Lactate provocation of panic attacks. I. Clinical and behavioral findings. Arch Gen Psychiatry 1984; 41:764-770.
10. de Beurs E, Buikhuisen M, Lange A et al. Continuous monitoring of panic. Acta Psych Scand 1994; 90:38-45.
11. Basoglu M, Marks IM, Sengun S. A prospective study of panic and anxiety in agoraphobia with panic disorder. Br J Psychiatry 1992; 160:57-64.
12. Price LH, Goddard AW, Barr LC et al. Pharmacological challenges in anxiety disorders. In: Bloom FE, Kupfer DJ, eds. Psychopharmacology: the fourth generation of progress. New York: Raven Press, 1995:1311-1323.
13. Nutt DJ, Glue P. Clinical pharmacology of anxiolytics and antidepressants: a psychopharmacological perspective. Pharmacol Ther 1989; 44:309-334.
14. Thomas N. Noradrenergic effects of various antidepressant therapies as studied by in vivo microdialysis. Ph.D. thesis, 1991; University of Reading, UK.
15. Charney DS, Heninger GR. Noradrenergic function and the mechanism of action of antianxiety treatment. II. The effect of long-term imipramine treatment. Arch Gen Psychiatry 1985; 42:473-481.

16. Charney DS, Price LH, Heninger GR. Desipramine-yohimbine combination treatment of refractory depression. Arch Gen Psychiatry 1986; 43:1155-1161.
17. Schmauss M, Laakman G, Dieterle D. Effects of $\alpha_2$-receptor blockade in addition to tricyclic antidepressants in therapy-resistant depression. J Clin Psychopharmacol 1988; 8:108-111.
18. Papp LA, Klein DF, Gorman JM. Carbon dioxide hypersensitivity, hyperventilation and panic disorder. Am J Psychiatry 1993; 150:1149-1157.
19. Gorman JM, Goetz RR, Dillon D. Sodium D-lactate infusion in panic disorder patients. Neuropsychopharmacol 1990; 3:181-189.
20. Anderson WC, Wetzler S. Five percent carbon dioxide challenge: valid analogue and marker of panic disorder? Biol Psychiatry 1990; 27:689-701.
21. Papp LA, Klein DF, Martinez J et al. Diagnostic and substance specificity of carbon-dioxide-induced panic. Am J Psychiatry 1993; 150:250-257.
22. Klein DF. False suffocation alarms, spontaneous panics, and related conditions. Arch Gen Psychiatry 1993; 50:306-317.
23. Hibbert GA, Pilsbury D. Hyperventilation: is it a cause of panic attacks? Br J Psychiatry 1989; 155:687-689.
24. Margraf J. Hyperventilation and panic disorder: a psychophysiological connection. Adv Behav Res Ther 1993; 15:49-74.
25. Salkovskis PM, Clark DM. Affective responses to hyperventilation: a test of the cognitive model of panic. Behav Res Ther 1990; 28:51-61.
26. van den Hout MA, Boek C, van der Molen GM et al. Rebreathing to cope with hyperventilation: experimental tests of the paper bag method. J Behav Med 1988; 11:303-310.
27. Charney DS, Bremner DJ, Redmond DE. Noradrenergic neural substrates for anxiety and fear: clinical associations based on preclinical research. In: Bloom FE, Kupfer DJ, eds. Psychopharmacology: the fourth generation of progress. New York: Raven Press, 1995:387-395.
28. Aston-Jones G, Rajkowski J, Kubiak P et al. Locus ceruleus neurons in the monkey are selectively activated by attended stimuli in a vigilance task. J Neurosci 1994; 14:4467-4480.
29. Rajkowski J, Kubiak P, Aston-Jones G. Activity of locus ceruleus neurones in behaving monkeys varies with focused attention: short- and long-term changes. Soc Neurosci Abstr 1992; 18:538.
30. Woods SW, Charney DS. Biological responses to panic anxiety elicited by nonpharmacologic means. In: Ballenger JC, ed. Neurobiology of panic disorder. New York: Wiley-Liss, 1990:205-217.
31. Ko GN, Elsworth JD, Roth RH et al. Panic-induced elevation of plasma MHPG levels in phobic-anxious patients: effects of clonidine and imipramine. Arch Gen Psychiatry 1983; 40:425-430.
32. Nesse RM, Curtis GC, Thyer BA et al. Endocrine and cardio-

vascular responses during phobic anxiety. Psychosom Med 1985; 47:320-332.
33. Yeragani VK, Srinivasan K, Balon R. Lactate sensitivity and cardiac cholinergic function in panic disorder. Am J Psychiatry 1994; 151:1226-1228
34. Dimsdale JE, Ziegler MG. What do plasma and urinary measures of catecholamines tell us about human responses to stressors? Circulation 1991; 83 (suppl II):36-42.
35. Kosten TR, Mason JW, Giller EL et al. Sustained urinary norepinephrine and epinephrine elevation in posttraumatic stress disorder. Psychoneuroendocrinol 1987; 12:13-20.
36. Yehuda R, Southwick SM, Giller EL. Urinary catecholamine excretion and severity of post-traumatic stress disorder in Vietnam combat veterans. J Nerv Ment Dis 1992; 180:321-325.
37. Rainey JM, Pohl RB, Williams M et al. A comparison of lactate and isoproterenol anxiety states. Psychopathology 1984; 17 (suppl.1):74-82.
38. Nesse RM, Cameron OG, Curtis GC et al. Adrenergic function in patients with panic anxiety. Arch Gen Psychiatry 1984; 41:771-776.
39. Charney DS, Woods SW, Krystal JH et al. Noradrenergic neuronal dysregulation in panic disorder: the effects of intravenous yohimbine and clonidine in panic disorder. Acta Psych Scand 1992; 86:273-282.
40. Albus M, Zahn TP, Breier A. Anxiogenic properties of yohimbine. I. Behavioral, physiological and biochemical measures. Eur Arch Psychiatry Clin Neurosci 1992; 241:337-344.
41. Yeragani VK, Berger R, Pohl R et al. Effects of yohimbine on heart rate variability in panic disorder patients and normal controls: a study of power spectral analysis of heart rate. J Cardiovasc Pharmacol 1992; 20:609-618.
42. Coupland NJ, Glue P, Nutt DJ. Challenge tests: assessment of the noradrenergic and GABA systems in depression and anxiety disorders. Mol Aspects Med 1992; 13:221-247.
43. Tancer ME, Stein MB, Uhde TW. Growth hormone responses to clonidine and growth hormone releasing factor in patients with panic disorder. Am J Psychiatry 1993; 150:336-337.
44. Holsboer F. Neuroendocrinology of mood disorders. In: Bloom FE, Kupfer DJ, eds. Psychopharmacology: the fourth generation of progress. New York: Raven Press 1995; 957-969.
45. Southwick SM, Krystal JH, Morgan CA et al. Abnormal noradrenergic function in posttraumatic stress disorder. Arch Gen Psychiatry 1993; 50:266-274.
46. Charney DS, Woods SW, Heninger GR. Noradrenergic function in generalized anxiety disorder: effects of yohimbine in healthy subjects and patients with generalized anxiety disorder. Psychiatry Res 1989; 27:173-182.
47. Rasmussen SA, Goodman WK, Woods SW et al. Effects of

yohimbine in obsessive compulsive disorder. Psychopharmacol 1987; 93:308-313.
48. Charney DS, Heninger GR. Noradrenergic function and the mechanism of action of antianxiety treatment. I. The effect of long-term alprazolam. Arch Gen Psychiatry 1985; 42:458-467.
49. Goddard AW, Woods SW, Sholomskas DE et al. Effects of the serotonin reuptake inhibitor fluvoxamine on yohimbine-induced anxiety in panic disorder. Psychiatry Res 1993; 48:119-133.
50. Valentino RJ, Aston-Jones GS. Determinants of locus ceruleus discharge. In: Bloom FE, Kupfer DJ, eds. Psychopharmacology: the fourth generation of progress. New York: Raven Press, 1995:373-385.
51. Schittecatte M, Charles G, Machowski R et al. Reduced clonidine rapid eye movement sleep suppression in patients with primary major affective illness. Arch Gen Psychiatry 1992; 49:636-642.
52. Al-Damluji S. Adrenergic control of the secretion of anterior pituitary hormones. Balliere's Clin Endocrinol Metab 1993; 7:355-392.
53. Coplan JD, Liebowitz MR, Gorman JM et al. Noradrenergic function in panic disorder: effects of intravenous clonidine pretreatment on lactate induced panic. Biol Psychiatry 1992; 31:135-146.
54. Uhde TW, Stein MB, Vittone BJ et al. Behavioral and physiologic effects of short-term and long-term administration of clonidine in panic disorder. Arch Gen Psychiatry 1989; 46(2):170-177.
55. Nutt DJ. Benzodiazepine dependence: new insights from basic research. In: Hindmarch I, Beaumont G, Brandon S, et al, eds. Benzodiazepines: current concepts. Chichester: John Wiley & son, 1990:19-41.
56. Bruno JP, Miller JA. Inhibition of GABAergic transmission: interactions with other neurotransmitter systems. In: Sarter M, Nutt DJ, Lister RG, eds. Benzodiazepine receptor inverse agonists. New York: Wiley-Liss, 1995:41-82.
57. Clow A, Glover V, Sandler M et al. Increased urinary tribulin output in generalised anxiety disorder. Psychopharmacol 1992; 95:378-380.
58. Davidson J, Glover V, Clow A et al. Tribulin in post-traumatic stress disorder. Psychol Med 1988; 4:833-836.
59. Duka T, Dorow R. Human experimental psychopharmacology of benzodiazepine receptor inverse agonists and antagonists. In: Sarter M, Nutt DJ, Lister RG, eds. Benzodiazepine receptor inverse agonists. New York: Wiley-Liss, 1995:243-270.
60. Stephens DN, Turski L. Adaptive changes to chronic treatments with ligands for benzodiazepine receptors: sensitization to inverse agonists and tolerance to agonists. In: Sarter M, Nutt DJ, Lister R. eds. Benzodiazepine receptor inverse agonists. New York: Wiley-Liss, 1995:83-113.
61. Nutt DJ, Glue P, Lawson CW et al. Flumazenil provocation of panic attacks. Arch Gen Psychiatry 1990; 47:917-925.
62. Bernik M, Gorenstein C, Gentil V. Flumazenil-precipitated with-

drawal symptoms in chronic users of therapeutic doses of diazepam. J Psychopharmacol 1991; 5:215-219.
63. Roy-Byrne PP, Cowley DS, Greenblatt DJ et al. Reduced benzodiazepine sensitivity in panic disorder. Arch Gen Psychiatry 1990; 47:534-538.
64. Ball DM, Glue P, Wilson S et al. Pharmacology of saccadic eye movements in man (1): effects of the benzodiazepine receptor ligands midazolam and flumazenil. Psychopharmacology 1991; 105:361-367.
65. Roy-Byrne PP, Lewis N, Villacres E et al. Preliminary evidence of benzodiazepine subsensitivity in panic disorder. Biol Psychiatry 1989; 26:744-748.
66. Cowley DS, Roy-Byrne PP et al. Benzodiazepine sensitivity in panic disorder: effects of radant alprazolam treatment. Chronic Neuropsychopharmacol 1995; 12:147-157.
67. Woods SW, Charney DS, Silver JM et al. Behavioral, biochemical, and cardiovascular responses to the benzodiazepine receptor antagonist flumazenil in panic disorder. Psychiatry Res 1991; 36:115-127.
68. Bradwejn J, Koszycki D, Couetoux du Tertre A et al. Effects of flumazenil on cholecystokinin-tetrapeptide-induced panic symptoms in healthy volunteers. Psychopharmacol 1994; 114:257-261.
69. Jones AKP, Liyi Q, Cunningham VV et al. Endogenous opiate response to pain in rheumatoid arthritis and cortical and subcortical response to pain in normal volunteers using positron emission tomography. Int J Clin Pharm Res 1991; 11: 261-266.
70. Schlegel S, Steinart H, Bockisch A et al. Decreased benzodiazepine receptor binding in panic disorder measured by iomazenil-SPECT. Eur Arch Psychiatry Clin Neurosci 1994; 244:49-51.
71. Charney DS, Heninger GR. Serotonin function in panic disorders: the effect of intravenous tryptophan in healthy subjects and patients with panic disorder before and during alprazolam treatment. Arch Gen Psychiatry 1986; 43:1059-1065.
72. Westenberg HGM, den Boer JA. Serotonin function in panic disorder: effect of L-5-hydroxytryptophan in patients and controls. Psychopharmacol 1989; 98:283-285.
73. Goddard AW, Sholomskas DE, Walton KE et al. Effects of tryptophan depletion in panic disorder. Biol Psychiatry 1994; 36:775-777.
74. Kahn RS, Wetzler S, van Praag HM et al. Behavioral indications for serotonin receptor hypersensitivity in panic disorder. Psychiatry Res 1988; 25:101-104.
75. Targum SD, Marshall LE. Fenfluramine provocation of anxiety in patients with panic disorder. Psychiatry Res 1989; 28:295-306.
76. Charney DS, Woods SW, Goodman WK et al. Serotonin function in anxiety: II. Effects of the serotonin agonist mCPP in panic disorder patients and healthy subjects. Psychopharmacol 1987; 92:14-24.
77. Germine M, Goddard AW, Sholomskas DE et al. Response to m-

chlorophenylpiperazine (mCPP) in panic disorder patients and healthy subjects: influence of reduction in intravenous dosage. Psychiatry Res 1995; (in press)
78. Lesch KP, Wiesmann M, Hoh A et al. 5-HT1A receptor-effector system responsivity in panic disorder. Psychopharmacol 1992; 106:111-117.
79. Cowen PJ. Neuroendocrine probes for 5HT receptor subtypes. Eur Neuropsychopharmacol 1993;306-308.
80. Deakin JFW, Graeff FG, Guimaraes FS. 5-HT receptor subtypes and the modulation of aversion. In: Marsden CA, Heal DJ, eds. Central serotonin receptors and psychotropic drugs. Oxford: Blackwell Scientific, 1992:147-174.
81. den Boer JA, Westenberg JA. Serotonin function in panic disorder: a double blind placebo controlled study with fluvoxamine and ritanserin. Psychopharmacol 1990; 102:85-94.
82. Rapee R. Psychological factors in panic disorder. Adv Behav Res Ther 1993; 15:85-102.
83. Clark DM. A cognitive approach to panic. Behav Res Ther 1986; 24:461-470.
84. McNally RJ, Lorenz M. Anxiety sensitivity in agoraphobics. J Behav Ther Exp Psychiatry 1987; 18:3-11.
85. Ackerman SH, Sachar EJ. The lactate theory of anxiety: a review and re-evaluation. Psychosom Med 1974; 36:69-81.
86. Rachman S, Levitt K, Lopatka C. Panic: the link between cognitions and bodily symptoms. Behav Res Ther 1987; 25:411-423.
87. Wilson KG, Sandler LS, Asmundson GJG. Fearful and non-fearful panic attacks in a student population. Behav Res Ther 1993; 31:407-411.
88. Costello CG. Problems in recent tests of two cognitive theories of panic. Behav Res Ther 1992; 30:1-5.
89. Street LL, Craske MG, Barlow DH. Sensations, cognitions and the perception of cues associated with expected and unexpected panic attacks. Behav Res Ther 1989; 27:189-198.
90. Thyer BA, Himle J. Phobic anxiety and panic anxiety: how do they differ?. J Anx Dis 1987; 1:59-67.
91. McNally RJ, Foa EB. Cognition and agoraphobia: bias in the interpretation of threat. Cog Ther Res 1987; 11:567-581.
92. Harvey JM, Richards JC, Dziadosz T et al. Misinterpretation of ambiguous stimuli in panic disorder. Cog Ther Res 1993; 17:235-248.
93. Clark DM, Salkovskis PM, Gelder MG et al. Tests of a cognitive theory of panic. In: Hand I, Wittchen HU, eds. Panic and phobias II. New York: Springer-Verlag, 1988.
94. Salkovskis PM, Clark DM, Hackmann A. Treatment of panic attacks using cognitive therapy without exposure or breathing retraining. Behav Res Ther 1991; 29:161-166.
95. Weinstein A, Neal L, Lillywhite A et al. Cognitive dysfunction in

panic disorder and post-traumatic stress disorder. Br J Clin Psychol, 1995 (in press).
96. Mellman TA, Uhde TW. Electroencephalographic sleep in panic disorder. Arch Gen Psychiatry 1989; 46:178-184.
97. Seligman MEP. Competing theories of panic. In: Rachman S, Maser JD, eds. Panic: psychological perspectives. New Jersey: Erlbaum, 1988.
98. Salkovskis PM. The importance of behaviour in the maintenance of anxiety and panic: a cognitive account. Behav Psychother 1991; 19:6-19.
99. Hibbert GA. Ideational components of anxiety: their origin and content. Br J Psychiatry 1984; 144:618-624.
100. Noyes R, Anderson DJ, Clancy J et al. Diazepam and propranolol in panic disorder and agoraphobia. Arch Gen Psychiatry 1984; 41:287-292.
101. Mavissakalian MR, Perel JM. Imipramine dose-response relationship in panic disorder with agoraphobia. Arch Gen Psychiatry 1989; 46:127-131.
102. Shear MK, Pilkonis PA, Cloitre M et al. Cognitive behavioral treatment compared with nonprescriptive treatment of panic disorder. Arch Gen Psychiatry 1994; 51:395-401.
103. Clark DM, Salkowskis PM, Hackmann A et al. A comparison of cognitive therapy, applied relaxation and imipramine in the treatment of panic disorder. Br J Psychiatry 1994; 164:759-769.
104. Koszycki D, Cox BJ, Bradwejn J. Anxiety sensitivity and response to cholecystokinin tetrapeptide in healthy volunteers. Am J Psychiatry 1993; 150:1881-1883.
105. McNally RJ. Panic disorder: a critical analysis. Guilford, 1994.
106. Clark DM, Hemsley DR. The effects of hyperventilation; individual variability and its relationship to personality. J Behav Ther Exp Psychiatry 1982; 13:41-47.
107. van den Hout MA, Griez E. Cognitive factors in carbon dioxide therapy. J Psychosom Res 1982; 26:209-214.
108. van der Molen GM, van den Hout MA, Vroemen J et al. Cognitive determinants of lactate-induced anxiety. Behav Res Ther 1986; 24:677-680.
109. Rapee RM, Mattick R, Murrell E. Cognitive mediation in the affective component of spontaneous panic attacks. J Behav Ther Exp Psychiatry 1986; 17:243-253.
110. Rapee RM, Telfer LA, Barlow DH. The role of safety cues in mediating the response to inhalation of $CO_2$ in agoraphobics. Behav Res Ther 1991; 29:353-355.
111. Clark DM. Cognitive mediation of panic attacks induced by biological challenge tests. Adv Behav Res Ther 1993; 15:75-84.
112. Bradwejn J, Koszycki D. Comparison of $CO_2$-induced panic attacks with cholecystokinin-induced panic attacks in panic disorder. Prog Neuropsychopharmacol Biol Psychiatry 1991; 15:237-239.
113. Bradwejn J, Koszycki D. Dose ranging study of the effect of

CCK4 in healthy volunteers. J Psychiatry Neurosci 1991; 16:260-264.
114. Bradwejn J, Koszycki D, Annable L et al. A dose-ranging study of the behavioral and cardiovascular effects of CCK-tetrapeptide in panic disorder. Biol Psychiatry 1992; 32:903-912.
115. Donnell CD, McNally RJ. Anxiety sensitivity and history of panic as predictors of response to hyperventilation. Behav Res Ther 1989; 27:325-332.
116. Koszycki D, Zacharko RM, Bradwejn J. The relationship between personality variables and response to CCK-4 challenge in patients with panic disorder. Canadian College of Neuropsychopharmacology Meeting 1994; May 30-June 2: Montreal, Canada (abstract).
117. Ehlers A. Interoception and panic disorder. Adv Behav Res Ther 1993; 15:3-21.
118. Asmundson GJG, Sandler LS, Wilson LS et al. Panic attacks and interoceptive acuity for cardiac sensations. Behav Res Ther 1993; 31:193-197.
119. Barsky AJ, Cleary PD, Sarnie MK et al. Panic disorder, palpitations and the awareness of cardiac activity. J Nerv Ment Dis 1994; 182:63-71.
120. Ehlers A, Margraf J, Roth WT et al. Anxiety induced by false heart rate feedback in patients with panic disorder. Behav Res Ther 1988; 26:1-11.
121. Koenigsberg HW, Pollak CP, Fine J et al. Lactate sensitivity in sleeping panic disorder patients and healthy controls. Biol Psychiatry 1992; 32:539-542.
122. Koenigsberg HW, Pollak CP, Fine J et al. Cardiac and respiratory activity in panic disorder: effects of sleep and sleep lactate infusions. Am J Psychiatry 1994; 151:1148-1152.
123. Abelson JL, Nesse RM. Pentagastrin infusions in patients with panic disorder. I. Symptoms and cardiovascular responses. Biol Psychiatry 1994; 36:73-83.
124. van Megen HJGM, Westenberg HGM, den Boer JA et al. Pentagastrin induced panic attacks: enhanced sensitivity in panic disorder patients. Psychopharmacol 1994; 114:449-455.
125. Tancer ME, Stein MB, Uhde TW. Effects of TRH on blood pressure and heart rate in phobic and panic patients. Biol Psychiatry 1990; 27:781-783.
126. Goldstein RB, Weissman MM, Adams PB et al. Psychiatric disorders in relatives of probands with panic disorder and/or major depression. Arch Gen Psychiatry 1994; 51:383-394.
127. Glue P, Nutt DJ, Coupland NJ. Stress and psychiatric disorder: reconciling social and biological approaches. In: Stanford SC, Salmon P, eds. Stress: from synapse to syndrome. London: Academic Press, 1993:53-73.

CHAPTER 2

# Molecular Forms and Regional Distribution of Cholecystokinin in the Central Nervous System

**Jens F. Rehfeld and Finn Cilius Nielsen**

## INTRODUCTION

The recognition of widespread neuronal synthesis of bioactive peptides (neuropeptides) in the brain over the last decades has advanced basic neurobiology and biologically based psychiatry significantly. The neuropeptide concept is much broader than that covering the original small group of hypothalamic peptides regulating the release of pituitary hormones. Hence, neuropeptides constitute a large number of highly potent transmitters widely expressed in all regions of the central and peripheral nervous systems. Sometimes neuropeptides are co-synthesized and operate synergistically with the small so-called classical neurotransmitters like monoamines, acetylcholine and amino acids. Paradoxically, however, early primitive neurons such as those in coelenterates use peptide-transmitters rather than the "classical" small transmitters. Hence, neuropeptides are apparently the true classical or original transmitters. Possibly, neuropeptides are involved also in functions such as neuronal growth and metabolism (for reviews, see refs. 1 and 2).

Among the 50 or so known neuropeptide systems in the mammalian brain, one of the largest groups consist of hormones syn-

*Cholecystokinin and Anxiety: from Neuron to Behavior*, edited by Jacques Bradwejn and Eero Vasar. 

thesized also in endocrine cells of the digestive tract. This group, named gut-brain peptides were discovered in the brain in the late 1970s and 1980s. Since methods for analysis of gut-brain peptides then were already at hand in several gut hormone laboratories, studies of their expression and molecular forms in the central nervous system progressed rapidly. Although most gut peptides are expressed only in discrete brain areas and in low concentrations (for instance secretin, gastrin, insulin, glucagon and motilin), some are expressed in most regions of the central and peripheral nervous systems (cholecystokinin (CCK), neuropeptide Y, somatostatin and vasoactive intestinal polypeptide (VIP)). Among these, and also among peptides originally isolated from brain tissue (tachykinins, neurotensin, enkephalins and others) CCK peptides hold a unique position for several reasons. First, the total amount in the mammalian central nervous system is larger than that of any other known neuropeptide. Second, the distribution is unique, with the highest CCK expression in the cerebral cortex.[3-7] The amount and concentrations in the brain are significantly higher than those of neuropeptide Y[8,9] which by some has been claimed to be the major neuropeptide in the mammalian brain.[10,11] Third, CCK has aroused particular interest in psychiatric research, since CCK peptides may play a role in anxiety, schizophrenia and eating disorders. The interest in neuronal CCK has been emphasized by a number of international symposia in the last decade devoted exclusively to CCK and its receptors. Especially interested readers are referred to the proceedings from these meetings.[12-16]

In spite of substantial interest, several fundamental questions about neuronal CCK are still unanswered, as reflected in the often confusing literature. The problem may partly be attributable to ignorance of the extensive molecular heterogeneity of CCK in the brain. Therefore this review addresses the heterogeneity in terms of biogenesis and structure of the different transmitteractive CCK peptides in the central nervous system. As an introduction for neurobiologists and psychiatrists who may be less familiar with gut endocrinology, a short summary of the long history of CCK may be relevant.

## HISTORY OF CCK

In 1856 the excellent French physiologist, Claude Bernard, noted that hydrochloric acid in the duodenum stimulated bile flow.[17] Wertheimer later discovered that the response persisted after cutting of the vagal and sympathetic nerves.[18] On the basis of these results and his own experiments, Fleig in 1904 suggested that acid and soap in the duodenum released a specific "sapocrinine" that stimulated bile secretion and flow.[19] Further evidence for a specific hormonal mechanism for gallbladder emptying in dogs was provided by Ivy and Oldberg in 1928, who on that occasion coined the name cholecystokinin.[20] Later Ivy et al and Sandblom provided evidence for the existence of CCK also in man.[21,22]

Independent of the cholecystokinetic studies, Harper and Raper in 1943 discovered a substance from the duodenal mucosa that stimulated the release of enzymes from the pancreas.[23] They named this substance pancreozymin (PZ). Harper and associates later purified PZ partially to a preparation that could be used for studies of secretion from the pancreas.[24] CCK and PZ were considered to be separate hormones until Mutt and Jorpes in Stockholm in the late 1960s purified a cholecystokinetic peptide from hog intestines.[25-27] Mutt and Jorpes demonstrated that their tritriadecapeptide molecule (CCK-33) also contained the PZ activity of the intestinal extracts.[25] Since the CCK mechanism was discovered earlier than that of PZ, consensus has now been reached to use the acronym CCK instead of PZ or the awkward CCK-PZ.

The availability of pure porcine CCK-33 from Mutts laboratory and later synthetic C-terminal fragments, above all CCK-8 from Squibb in New Jersey[28,29] accelerated the research in two interrelated directions. First, structure-function studies of the pancreozymic and cholecystokinetic activity of CCK were performed at the cellular as well as the subcellular level (for review, see ref. 30). Second, the development of radioimmunoassays for CCK was attempted with the purpose of correlating the activity with the concentration and molecular nature of CCK in plasma (for review, see ref. 31).

The neuronal chapter of the CCK story began in 1975, when Vanderhaeghen et al reported that the vertebrate brain contained a small peptide reacting with gastrin antibodies.[32] This observation intrigued ours and two other laboratories interested in the molecular nature of gastrin. All laboratories subsequently reported that this gastrin-like peptide in the brain was CCK, which cross-reacted with gastrin antibodies due to the homologous C-terminal sequence of gastrin and CCK.[31,33-35] True gastrin peptides, are also present in the brain, but only in a few regions.[36] Notably CCK, in addition to central neurons, are expressed also in peripheral nerves; most abundantly so in the distal part of the gut.[4,37]

## BIOGENESIS OF CCK PEPTIDES IN NEURONS

CCK peptides are encoded by a single gene consisting of three exons. So far, alternative splicing of the CCK gene transcript has not been observed, and the mature mRNA encodes a preproCCK of 115 amino acids.[38,39] PreproCCK consists of a N-terminal signal sequence, the sequence containing the bioactive forms from CCK-83 to CCK-5 and finally a C-terminal flanking peptide. During translation preproCCK is truncated to proCCK by removal of the signal sequence, after which the different transmitteractive peptides are generated during the post-translational maturation of proCCK. The biosynthetic maturation is shown in Figure 2.1. It is expedient to distinguish different categories of preproCCK products. ProCCK is defined as the truncated preproCCK from which the signal sequence has been removed. Glycine-extended CCKs are a group of biosynthetic intermediates that are processed at the C-terminus and that constitute the immediate precursors of the mature transmitteractive CCK. Carboxyamidated CCKs are the bioactive CCK, in which the C-terminal phenylalanyl residue is α-carboxyamidated. Each category contains peptides of different chain length (see below).

The dynamics of cerebral CCK peptide synthesis have been studied in some detail.[40-43] Combination of these results with our general knowledge about peptide hormone synthesis (for reviews, see refs. 44 and 45) suggests the following picture of the biosynthetic pathway in CCK neurons: after translation of CCK mRNA in the rough endoplasmic reticulum and cotranslational truncation of the N-terminal pre- or signal sequence from preproCCK, intact proCCK is transported to the Golgi apparatus. As for other neuropeptides, the first post-translational modifications occur in the Golgi apparatus, where three tyrosyl residues (Tyr-77, -92, and -95) of proCCK are O-sulfated. The trypsin-like endoproteolytic cleavages at multiple monobasic and one dibasic processing site also begin in the trans-Golgi apparatus. Processing continues in small vesicles that carry the processing intermediates toward the axonal synapses, where the processing terminates in the synaptic vesicles. The last and decisive processing steps occur during storage and maturation in the synaptic vesicles. The vesicles contain

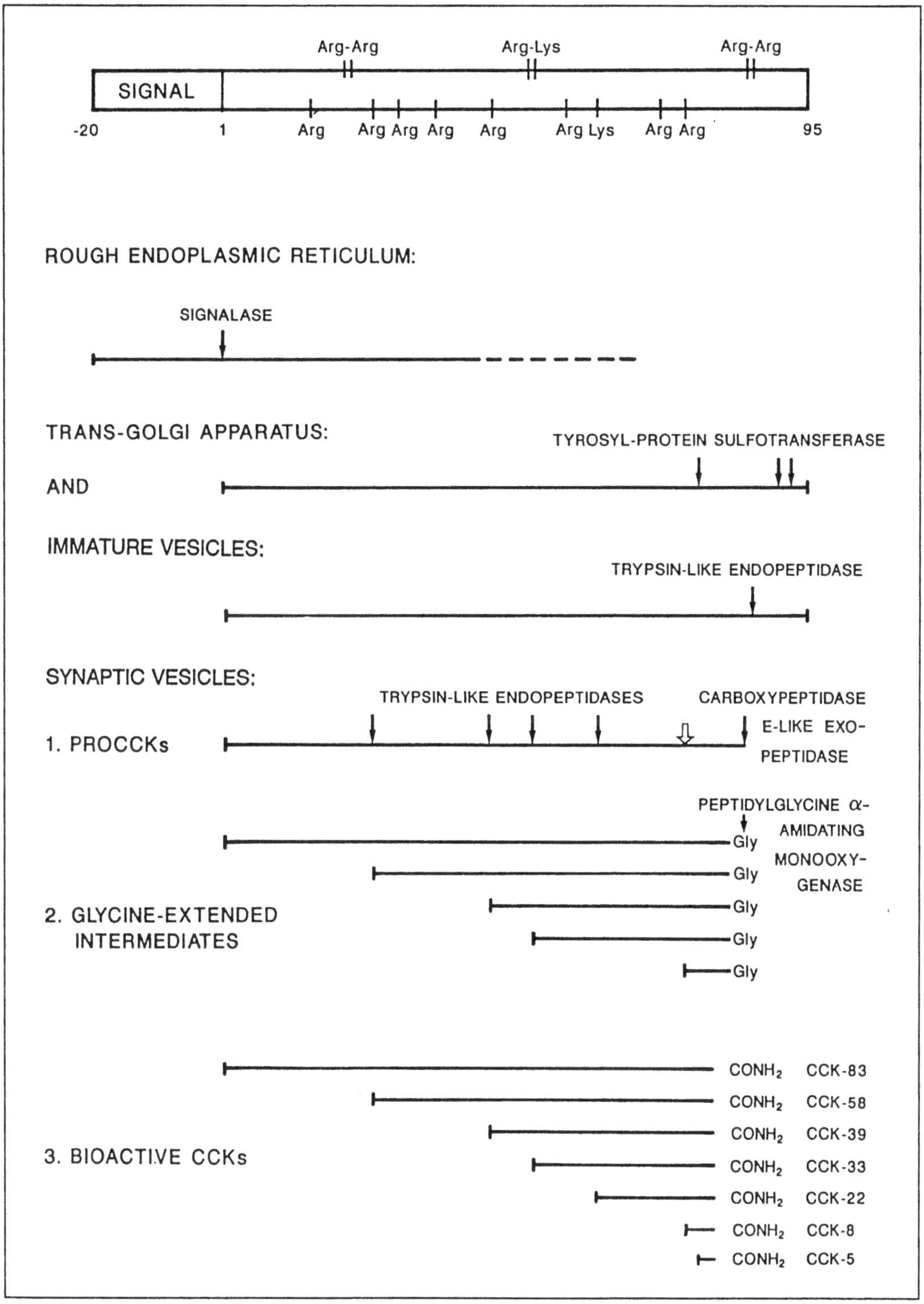


*Fig. 2.1. Schematic illustration of the biosynthesis cholecystokinin in neurons. The structure of preprocholecystokinin is illustrated as the bar in the upper part of the figure.*

the single precursor for the two enzymes necessary for amidation.[46-49] These enzymes remove glyoxylate from the immediate precursors, the glycine-extended CCKs, to complete the synthesis of bioactive α-carboxyamidated peptides (Fig. 2.1).

As a result of the elaborate biosynthetic pathway, CCK neurons release a heterogeneous mixture of proCCK products from the mature vesicles. A small percentage are non-amidated precursors. The amidated CCKs constitute a mixture of the longest possible bioactive product of proCCK,[50] in addition to medium-sized[3,26,51-53] and short CCKs.[43,54,55] The distribution and release patterns may vary among species.[56-58] However, irrespective of species, the predominant molecular form released from most neuronal tissues is CCK-8 in its O-sulfated and carboxyamidated form.

## MOLECULAR FORMS OF CCK IN THE BRAIN

We have tried to characterize the molecular picture of CCK peptides expressed in the cerebral cortex. In order to avoid species differences between the material and the assays, we used porcine, and when possible, human brains. Large quantities of the easily obtainable porcine cerebral cortex allowed characterization of components present in low quantities or having low immunoreactivity. The tissue was extracted sequentially in boiling water and dilute, cold acetic acid. The extractions and chromatographic fractionations were monitored with a library of assays specific for different sequences of proCCK. Moreover, assays specific for different sequences of the related gastrin and its precursor[59,60] ensured accurate detection of gastrin peptides when present. Some of the peptides were also purified and sequenced. The results showed that the brain indeed contains a multiplicity of different molecular forms.

### CARBOXYAMIDATED TRANSMITTERACTIVE FORMS

As already mentioned, carboxyamidated and tyrosine O-sulfated CCK-8 constitutes a major part of CCK in the cerebral cortex.[3,54,61,62] There are, however, three different forms of sulfated CCK-8 with different net charges resulting in a reproducible fractionation pattern on ion-exchange chromatography.[54,61] Only the last eluted form corresponds to native carboxyamidated and sulfated CCK-8. The sizes and ratio of the three forms vary from extract to extract. At present we know neither the nature of the subtle structural differences between the forms, nor whether the differences have biological significance. Perhaps they are merely extraction artifacts. Extensive boiling of synthetic CCK-8S converts part of the last eluted form (the native form) to the two less acidic forms, suggesting that oxidation of one or both methionine residues and perhaps of tryptophan may occur.[61] However, boiling of synthetic CCK-8 together with brain tissue apparently prevents such conversion (unpublished studies).

Nonsulfated CCK-8 is also present, although in concentrations of 1% or less of those of sulfated CCK-8[43] in neutral and dilute acetic acid extracts.[7,43] The small amounts are not simply an extraction artifact, since nonenzymatic desulfation occurs only under highly acid conditions.

In addition to CCK-8 there are substantial amounts of larger molecular weight forms containing the intact sulfated and carboxyamidated CCK-8 sequence at the C-terminus. By simple gel chromatography of extracts from the cerebral cortex four larger forms were distinguishable. A long form eluted in a position corresponding to CCK-83.[43] Another form eluted like CCK-58.[63] Two forms eluted like CCK-39 and CCK-33. The larger carboxyamidated forms elute after tryptic cleavage as carboxyamidated and O-sulfated CCK-8 and react fully with antisera specific for both the N-terminal and C-terminal sequence of CCK-8.[43]

The cerebral cortex also contains smaller carboxyamidated fragments of CCK-8. In addition to minimal amounts of sulfated CCK-7, there is a major component that elutes in the same position as CCK-5.[3,7,55,64] Purification of this component has shown that most is CCK-5, but that small amounts of CCK-4 are also present. Interestingly, CCK-4 and CCK-5 have almost the same affinity as CCK-8 to the CCK-B receptor, which is the predominating type of CCK-receptor in the brain.[65,66] The affinity to the CCK-B receptor contrasts with the poor binding to CCK-A receptors, which bind CCK-4 with a 10,000-fold lower potency than CCK-8.[67,68] CCK-A receptors, however, occur only sparsely in the brain. The high affinity of CCK-4 and -5 to the B-receptors suggests that the small C-terminal fragments per se may be neurotransmitters in the same way as CCK-4 has been suggested to act as transmitter for peripheral CCK nerves in the islets of Langerhans.[69] We do not know whether there is a cell-specific processing of proCCK in different neurons to particular molecular forms of carboxyamidated CCK.[43,70] The idea of neuron-specific processing is attractive, since it offers a new dimension of understanding to the ubiquitous expression of the CCK gene in the central nervous system. But unfortunately it is difficult to examine the idea.

### GLYCINE-EXTENDED FORMS

The amide donor for C-terminally amidated neuropeptides is glycine positioned next to the actual C-terminal residue of the active peptides in the precursor (Fig. 2.1). Glycine is usually followed by a pair of basic amino acids. Hence, proCCK products may occur extended by -Gly or the sequence -Gly-Arg-Arg- next to the C-terminal phenylalanine in the precursor,[38,39,71] (Fig. 2.1). To examine the occurrence of the precursor and C-terminally extended forms in the brain, we developed assays specific for glycine-extended CCK. Such assays can in combination with trypsin and carboxypeptidase-B treatment disclose the occurrence of both the gastrin precursor, the CCK precursor and forms with a free glycine at the C-terminus. In the cerebral cortex there is a small, but significant amount of glycine-extended CCKs present, indicating that proCCK is not processed completely to carboxyamidated CCKs. On the other hand we do not know whether glycine-extended CCKs as such have their own receptor, perhaps related to growth stimulation as suggested for glycine-extended gastrins.[72] Glycine-extended CCKs occur in forms corresponding to the carboxyamidated CCKs, in the adult brain in much lower concentrations, however.

### PROCCK AND INACTIVE N- AND C-TERMINAL FRAGMENTS OF PROCCK

In some water extracts of the cerebral cortex and especially of hippocampal areas with high concentrations of CCK cell bodies, there are also substantial amounts of a larger nonamidated form eluting like proCCK.[42]

As shown first by Eng et al[41] there are in the cerebral cortex also a number of N-terminal fragments of proCCK and the large molecular weight forms complementary to CCK-8. Whether these fragments are degraded and are without biological effects per se, or whether they have hitherto unknown effects unrelated to the effects of the well-known CCKs remains to be examined. We have, also found large quantities of these N-terminal fragments.[43] However, in contrast to Eng et al,[41] we found also substantial quantities of large molecular forms, CCK-33, -58, and even larger forms. The discrepancy can probably be explained by the use of different extraction techniques.

## REGIONAL DISTRIBUTION OF CCK IN THE BRAIN

Numerous CCK neurons are present in most parts of the brain with the exception of the cerebellum, in which CCK is poorly expressed.[41,73] Networks of high density were detected in the cerebral cortex including both neocortical regions and the entorhinal cortex. In the cortex, CCK peptides are present mainly in thin (terminal) nerve processes throughout all cortical layers, but slightly less in the molecular than in deeper layers. Most cortical CCK nerve processes are very thin. Therefore, the staining of CCK nerves produces a general light staining of the entire cortex.[4] When examined at higher magnifications, individual nerve processes could be discerned and numerous thicker, intensely immunoreactive nerves were also detected. The majority of cortical neurons do not express CCK. CCK nerve cell somas are, however, present in the neocortex. These somas usually occur in laminae II-IV and posses long processes, which run downwards from the brain surface. We have not found major regional differences in the distribution of CCK nerves in the neocortex. In most mammals, however, the enthorhinal cortex contains higher densities of CCK nerve fibers than the neocortex. In addition, the hippocampal formation as well as the amygdaloid nuclei contain numerous CCK neurons. In the lateral and caudal aspects of the hippocampus proper, the nerves are mainly confined to a band beneath the nerve cell layer, whereas in the medial and anterior aspects the nerves mingled with the nerve cell somas, which they innervate (pyramidal cells of CA1). Occasional CCK nerve cell somas are present in this layer, but they represent less than 1% of the hippocampal nerve cell body population. CCK nerve fibers were also detected in the dentate gyrus, where they seemed to mingle with some CCK nerve cell bodies. The amygdaloid area is also rich in CCK nerves. Networks of high densities were detected in the medial amygdaloid nucleus whereas the cortical amygdaloid nucleus contained a less dense network than the other nuclei networks of medium to low densities. The caudate nucleus and the putamen contained moderately developed networks of CCK, which were more dense at rostral than

***Table 2.1. Cerebral expression of carboxyamidated cholecystokinin peptides in the central nervous system of guinea pigs***

| Region | pmol CCK/g tissue |
|---|---|
| Telecephalon | |
| Frontal cortex | 214 |
| Parietal cortex | 145 |
| Temporal cortex (including the amygdaloid nuclei) | 280 |
| Occipital cortex | 76 |
| Olfactory lobe | 231 |
| Hippocampus | 55 |
| Caudate nucleus and putamen | 167 |
| Diencephalon | |
| Thalamus | 45 |
| Hypothalamus | 74 |
| Metencephalon | |
| Cerebellum | 1 |
| Pons (rostral) | 113 |
| Myelencephalon | |
| Medulla oblongata (rostral) | 25 |
| Medulla oblongata (caudal) | 27 |
| Spinal cord | |
| Medulla spinalis cerv. | 22 |
| Medulla spinalis thor. | 28 |
| Medulla spinalis lumb. | 34 |
| Medulla spinalis sacr. | 63 |
| Spinal ganglion | 22 |

at caudal levels. Also the hypothalamus is rich in CCK neurons. Networks of particularly high densities were detected in the medial preoptic area and also the periventricular and lateral hypothalamic areas contained much CCK. Few nerves are in addition scattered throughout the internal and external zones of the median eminence. Other regions containing moderately developed networks of CCK neurons include the lateral septal nuclei, the periaqueductal grey of the midbrain and the area postrema. Structures that contain only few or no CCK neurons are the cerebellum, the corpus callosum, the internal capsule, the anterior and posterior commissures and the subcommissural organ. The results of immunochemical quantitation of carboxyamidated CCK peptides expressed in different regions of the central nervous system is shown in Table 2.1.

***Table 2.2. Extracerebral expression of cholecystokinin peptides in porcine tissue***

| **Tissue (pmol/g tissue)** | **Total translation product percentage** | **Precursor** |
|---|---|---|
| Gastrointestinal tract | | |
| Duodenal mucosa | 200 | 5 |
| Jejunal mucosa | 250 | 20 |
| Ileal mucosa | 20 | 50 |
| Colonic mucosa | 5 | 50 |
| Neuroendocrine tissue | | |
| Adenohypophysis | 25 | 100 |
| Neurohypophysis | 20 | 10 |
| Thyroid gland | 2 | 20 |
| Adrenal medulla | 1 | 50 |
| Genital tract | | |
| Testicles | 5 | 80 |
| Spermatozoa | – | – |

## EXTRACEREBRAL EXPRESSION OF PROCCK

CCK is also produced in several cell types other than cerebral CCK neurons — above all in the endocrine intestinal I-cells. CCK peptides are also widely expressed in peripheral neurons, primarily in the intestinal tract, but also in the genitourinary tract and elsewhere.[4,37] In addition, low-level expression has been found in pituitary corticotrophs,[74,75] in thyroid C cells,[76] in the adrenal medulla,[77] in the bronchial mucosa[78] and in spermatogenic cells of certain mammals.[79] Quantitative measures of the tissue-specific extracerebral expression are presented in Table 2.2.

## CCK PEPTIDES AS NEUROTRANSMITTERS

Neurotransmitters are defined as chemicals released by nerve endings that alter the activities of adjacent neurons or organs. It has been suggested that neuropeptides without clear-cut transmitter effects should be named neuromodulators. Such labeling is, however, "a semantic masking of our ignorance of synaptic mechanisms."[80] Full examination of transmitter status has been performed only for the predominant CCK peptide in the nervous system, i.e., tyrosyl-sulfated and carboxyamidated CCK-8. Present evidence suggests, however, that all carboxyamidated

CCK peptides can act as transmitters. For CCK-8 the following criteria have been fulfilled:

### Localization in neurons

CCK-8 is localized in an extensive network of nerves in the CNS, in cell bodies as well as nerve endings.[4] Also, in the gut, pancreas, urinary bladder and in peripheral somatic nerves CCK-8 has been located in neurons.[4,69,81,82]

### Concentration in nerve terminals

Specific immunocytochemistry has shown that CCK peptides are concentrated in nerve terminals. This impression has been substantiated by subcellular fractionation of brain tissue, which demonstrated a fourfold greater concentration of CCK-8-like peptides in synaptosomes and synaptic vesicles.[83,84]

### Demonstration of synthesis

Intraventricular pulse administration of [$^{35}$S]-methionine in anesthetized adult rat brains has revealed an extensive in vivo biosynthesis of CCK in the cerebral cortex.[40] Radioactive methionine was first incorporated in large molecular forms of CCK (-83 and -58), then in CCK-33-like material and finally in CCK-8 and CCK-5-like peptides.

Chasing with unlabeled methionine revealed the large forms and CCK-8, whereas the relationship with the CCK-5-like material was less obvious.[40]

### Release by depolarization

Superfusion of brain slices with potassium chloride induced a calcium-dependent release of CCK-8-like material in rats, a feature typical of neurotransmitters.[84,85]

### Mimicking of transmitter effect

Application of small amounts of CCK-8 (fmol/s) to the postsynaptic membrane strongly excited hippocampal neurons in the rat.[86] Also worth mentioning is that perfusion with picomolar concentrations of CCK-4 in the isolated, perfused pig pancreas induced a dramatic release of insulin and glucagon in accordance with the innervation of pancreatic islets with CCK-containing terminals.[69]

### INACTIVATION

Whether CCK-8 is inactivated by degradation and/or by reuptake from the synaptic cleft has not yet been fully determined. The rapid de novo synthesis of CCK-8 makes reuptake less likely.[40] Moreover, evidence of CCK-8-degrading enzymes has now been found in the rat brain.[42,87]

### INTERFERENCE BY ANTAGONISTS

Final acceptance of CCK-8 as a neurotransmitter requires demonstration of interference or abolition of the transmitter effect by suitable receptor antagonists in a way similar to the interference exerted by naloxone on the activity of opioids. Already years ago, the low-affinity CCK-receptor antagonist, proglumide fulfilled this criterion.[88,89] The recent highly specific high-affinity antagonists have removed any doubt (for review, see ref. 15).

## PROBLEMS IN THE IDENTIFICATION OF NEURONAL CCK PEPTIDES

### SPECIES DIFFERENCES

It is well known that CCK peptides display a number of species variations.

First, the amino acid sequence of CCK varies. Although proper sequence data on peptide isolated from the brain still are limited to a few mammalian species,[41,54] the C-terminus of CCK seems very similar in most mammals. In contrast, there are a few but essential amino acid substitutions in the N-termini of CCK-33 and -58 between, e.g., pig,[27,41] dog,[63] rat[39] and man,[39] rendering quantitation and histochemical demonstration by immunotechniques inconsistent with several antisera raised against porcine CCK. Second, the concentration of CCK in different brain regions varies considerably (Table 2.1). A region such as the neurohypophysis illustrates the species variation. Whereas there is no CCK in the porcine and feline neurohypophysis, bovine and murine neuropituitaries contain the highest CNS concentrations.

Third, the size of the brain or rather the amount of tissue available plays a role. It is of course easy to obtain sufficient amounts of tissue from large domestic animals such as pigs, cows and sheep. The amount of tissue becomes crucial with certain molecular forms, which are present either in low concentrations or display low reactivity with available assays. Thus, studies that fail to discover certain molecular forms in small tissue samples, typically from brains of small rodents, may well be false negative.

EXTRACTION PROBLEMS

A major problem in the study of brain CCK and a (misleading) reason for considering neuronal CCK as being synonymous with only CCK-8 is on one hand that different extraction techniques have been used in the different laboratories,[3,5,6,32,33,41,62,90-94] and on the other hand that the different molecular forms require different extraction techniques. As a consequence of factors such as the solubility of the peptides, their chemical stability and the physical properties of the tissue to be extracted, the extract probably never contains all the relevant peptides present in the tissue. Moreover, structural changes, even small species differences, may change the yield of certain extraction techniques. Finally, some of the extraction techniques employed are so crude that they modify the peptide structure. In particular, amino acid side groups such as (in the actual case) tyrosine-*O*-sulfate are easily removed by extraction at extreme pH values.

Studies on the effects of different extraction procedures on the yield and structure of CCK indicate that only three methods can be used to combine the highest yield with minimal damage to the peptide. Careful boiling in water at neutral pH extracts all the small forms (CCK-8, -5 and -4) and in certain regions also a large form corresponding to CCK-83. Boiling of pure CCK-8 itself in the absence of tissue, but otherwise under similar conditions, changes the net charge of the molecule, perhaps by sulfoxidation of the two methionine residues or by some other modifications. However, in the presence of fresh brain tissue this effect is minimal.

Ice-cold 90% methanol/10% water yields a near-maximal extraction of CCK-8, sulfated and nonsulfated, essentially without structural modifications. However, forms of molecular weights lower than CCK-8 are only partly extracted and larger forms not extracted at all.[61]

In ice-cold 0.5 M acetic acid most large and medium sized forms (CCK-83, -58, -39, and -33) as well as larger N-terminal fragments are released from the tissue. Only a minor fraction of CCK-8 and smaller forms are extracted in acetic acid. Cold, dilute acetic acid is acceptable — in contrast to dilute HCl used by some laboratories.[6,41,92] Acidic extraction has to be so mild that the peptides are neither desulfated nor modified in other respects.

## ASSAY PROBLEMS

Although CCK molecules from the brain have been examined with respect both to bioactivity[54,90] and receptor binding,[65,66,95-97] neither bioassays nor radioreceptor assays have proven adequate for extensive quantitation and characterization of CCK peptides in extracts and chromatographical fractions. For all practical purposes the radioimmunoassay (RIA) technique is most convenient. This is, however, not to say that RIA is without problems. On the contrary, the structural homology with gastrin, which posed problems in the first study of brain CCK,[32] and the molecular heterogeneity of cerebral CCK (Fig. 2.1) have caused considerable difficulties and controversies in the interpretation of the RIA data.

Since it has been difficult to develop truly specific antibodies for carboxyamidated CCK[98] (for review, see ref. 31 and 99) most laboratories have simply used C-terminal-directed gastrin immunoassays, which cross-react to varying extent with CCK molecules.[5,6,32-34,93,100] In so doing, mistakes may occur such as overlooking gastrin in brain regions containing much CCK.[101-103] Also, N-terminal fragments[41] and C-terminal extended forms escape detection by such assays. Finally, the use of the C-terminal, cross-reacting assays requires careful calibration with both N-terminal extended forms and C-terminal fragments of CCK-8. In other words, theses immunoassays generally measure both larger molecular CCKs (-83, -58, -33 and -22) and short fragments (CCK-5 and -4) with reduced binding affinity. To characterize brain CCK fully it is necessary to rely not on a single assay, but on a library of assays specific for different sequences of both active peptides and their precursors.[98,104] Such libraries of sequence-specific antisera are necessary also for immunocytochemical characterization of the peptide-producing neurons.

## FRACTIONATION PROBLEMS

Today several powerful fractionation techniques are available for purification and identification of peptides in tissue extracts. Many laboratories use gel, ion-exchange and reversed phase liquid chromatography in their CCK-studies. The high resolution capacity of reversed phase chromatography is of course a great advantage, but poses problems with minor modifications of the peptides (i.e., extraction and the species problem). Hence, calibration of high pressure chromatography columns with porcine CCK- and gastrin peptides may be misleading in studies of brain extracts for non-porcine species. A single amino acid substitution may change the elution position significantly and make the interpretation of data incorrect.

## MOLECULAR FORMS OF GASTRIN IN THE BRAIN

The antral hormone gastrin is closely related to CCK. Moreover, carboxyamidated gastrin peptides bind to CCK-B receptors with the same affinity as carboxyamidated CCK peptides. It is therefore relevant to consider also the expression and synthesis of gastrin peptides in the brain. Up to now, true gastrin molecules have been found in only three areas of the central nervous system. There are gastrins in small but significant amounts in the oxytocinergic hypothalamo-neurohypophyseal neurons of all mammalian species examined so far.[36,75,105] The neuropituitary gastrins are the same forms as found in the antral gastrin cells, i.e., gastrin-71, gastrin-34 and gastrin-17. In the pig pituitary, which we have examined most extensively, half of these three gastrin forms are tyrosine-*O*-sulfated, in the same way as CCK.[75] The function of the neuropituitary gastrins is unknown. However, the permanent presence in all mammals suggests that they are there for some reason. As mentioned above, the occurrence of CCK in the pituitary varies considerably.

The second gastrin-containing region in the brain is medulla oblongata in the cat, in which we occasionally have found small amounts of gastrin-34 and larger amounts of gastrin-17.[82] Whether this gastrin is related to the occasional occurrence of gastrin in vagal nerve fibers[81,82,106] remains to be shown.

The third part of the brain expressing true gastrin peptides is the cerebellum.[107] The concentrations are low, only a few pmol per g tissue. The role of cerebellar gastrin, which occurs mainly as gastrin-17, is unknown. But presumably the gastrins also act as transmitters via postsynaptic binding to the abundant CCK-B receptors. The small amounts of gastrin in the human CSF (cerebrospinal fluid)[108] presumably originates from the three regions in the brain, since there is no correlation with plasma gastrin concentrations.[109]

## CONCLUSION

It is now two decades since CCK and gastrin were found in central and peripheral neurons. Today we know that CCK is the most abundant peptide system in the brain. Although massive amounts of information have accumulated in the meantime and motivated extensive research, our ignorance about both the molecular and cellular structure and function of CCK in the brain is still substantial. When it comes to their role in CNS diseases, our ignorance is even more obvious. As a basis for further progress, precise information about the molecular structure of CCK peptides in the different brain regions is therefore a condition *sine qua non.*

### ACKNOWLEDGMENT

The skillful secretarial assistance of Gitte Runge Hansen is gratefully acknowledged. This review is based on studies, which have been supported by grants from the Danish MRC and the Danish Biotechnology program for Signal Peptide Research.

### REFERENCES

1. Krieger DT. Brain peptides: what, where and why? Science 1983; 222:975-985.
2. Krieger DT, Brownstein MJ, Martin JB., eds. John Wiley & Son, New York. Brain Peptides 1984.
3. Rehfeld JF. Immunochemical studies on cholecystokinin. II. Distribution and molecular heterogeneity in the central nervous system and small intestine of man and hog. J Biol Chem 1978; 253:4022-4030.
4. Larsson L-I, Rehfeld JF. Localization and molecular heterogeneity of cholecystokinin in the central and peripheral nervous system. Brain Res 1979; 165:201-218.
5. Lamers CB, Morley JE, Poitras P. et al. Immunological and biological studies on cholecystokinin in rat brain. Am J Physiol 1980; 239:E232-E235.
6. Beinfeld MC, Palkovits M. Distribution of cholecystokinin in the hypothalamus and limbic system of the rat. Neuropeptides 1981; 2:123-129.
7. Marley PD, Rehfeld JF, Emson PC. Distribution and chromatographic characterisation of gastrin and cholecystokinin in the rat central nervous system. J Neurochem 1984; 42:1523-1535.
8. Tatemoto K, Carlquist M, Mutt V. Neuropeptide Y - a novel brain peptide with structural similarities to peptide YY and pancreatic polypeptide. Nature 1982; 296:659-660.

9. Tatemoto K. Neuropeptide Y: complete amino acid sequence of the brain peptide. Proc Natl Acad Sci USA 1982; 79:5485-5489.
10. Allen YS, Adrian TE, Allen JM et al. Neuropeptide Y distribution in the rat brain. Science 1983; 221:877-879.
11. Adrian TE, Allen JM, Bloom SR et al. Neuropeptide Y distribution in human brain. Nature 1983; 306:584-586.
12. de Belleroche J, Dockray GJ., eds. Cholecystokinin (CCK) in the nervous system: current developments in neuropeptide research. Ellis Horwood Health Science Series, Chichester, UK 1984:1-132.
13. Crawley J, Vanderhaeghen J-J. eds. Neuronal cholecystokinin. Ann NY Acad Sci 1985; 448:1-686.
14. Dockray GJ, Woodruff G., eds. The neuropeptide cholecystokinin. Ellis Horwood Ltd., Chichester UK, 1989:1-244.
15. Dourish CT, Cooper SJ, Iversen SD et al., eds. Multiple cholecystokinin receptors in the CNS. Oxford University Press 1992:1-541.
16. Reeve JR, Eysselein V, Solomon T et al., eds. Cholecystokinin. Ann NY Acad Sci 1994; 713:1-458.
17. Bernard C. Lecons de Physiologie Experimental 1856; 2:429.
18. Wertheimer E. De l'action des acides et du chloral sur la sécretion biliaire. Comp Rend Soc Biol 1903; 55:286-287.
19. Fleig C. Du mode d'action des exitants chimique des glandes digestives. Arch Int Physiol 1904; 1:286-346.
20. Ivy AC, Oldberg E. Hormone mechanism for gallbladder contraction and evacuation. Am J Physiol 1928; 86:599-613.
21. Ivy AC, Drewyer GE, Orndoff BH. Effect of cholecystokinin on human gallbladder. Endocrinology 1930; 14:343-348.
22. Sandblom P. Function of human gallbladder studies in connection with blood transfusions and after stomach operations. Acta Radiol 1933; 14:249-258.
23. Harper AA, Raper HS. Pancreozymin, a stimulant of the secretion of pancreatic enzymes in extracts of the small intestine. J Physiol 1943; 102:115-125.
24. Crick J, Harper AA, Raper HS. On the preparation of secretin and pancreozymin. J Physiol 1949; 110, 367-376.
25. Mutt V, Jorpes JE. Isolation of aspartyl-phenylalanine amide from cholecystokinin-pancreozymin. Biochem Biophys Res Commun 1966; 26:392-397.
26. Mutt V, Jorpes JE. Structure of porcine cholecystokinin-pancreozymin. I. Cleavage with thrombin and with trypsin. Eur J Biochem 1968; 6:156-162.
27. Mutt V, Jorpes JE. Hormonal polypeptides of the upper intestine. Biochem J 1971; 125:57P-58P.
28. Ondetti MA, Pluscec J, Sabo EF et al. Synthesis of cholecystokinin-pancreozymin. I. The C-terminal dodecapeptide. J Am Chem Soc 1970; 92:195-216.

29. Bodanzsky M, Chaturvedi N, Hudson D et al. Cholecystokinin-pancreozymin. I. The synthesis of peptides corresponding to the N-terminal sequence. J Org Chem 1972; 37:2303-2312.
30. Jorpes JE, Mutt V. Secretin and cholecystokinin (CCK). In eds. Jorpes JE and Mutt V. Handbook of Experimental Pharmacology Vol. XXXIV: Secretin, Cholecystokinin, Pancreozymin and Gastrin. Springer Verlag, Berlin 1973:1-179.
31. Rehfeld JF. How to measure cholecystokinin in plasma? Gastroenterology 1984; 87:434-436.
32. Vanderhaeghen JJ, Signeau JC, Gepts W. New peptide in the vertebrate CNS reacting with gastrin antibodies. Nature 1975; 257:604-605.
33. Dockray GJ. Immunochemical evidence of cholecystokinin-like peptides in brain. Nature 1976; 264: 568-570.
34. Müller JE, Straus E, Yalow RS. Cholecystokinin and its COOH-terminal octapeptide in the pig brain. Proc Natl Acad Sci USA 1977; 74:3035-3037.
35. Rehfeld JF. Gastrins and cholecystokinin in brain and gut. Acta Pharmacol Toxicol 1977; 24:44.
36. Rehfeld JF. Localization of gastrin to neuro- and adenohypophysis. Nature 1978; 271:771-773.
37. Larsson L-I, Rehfeld JF. Peptidergic and adrenergic innervation of pancreatic ganglia. Scand J Gastroenterol 1979; 14:433-437.
38. Deschenes RJ, Lorenz LJ, Haun RS et al. Cloning and sequence analysis of a cDNA encoding rat preprocholecystokinin. Proc Natl Acad Sci USA 1984; 81:726-730.
39. Takahashi Y, Kato K, Hayashizaki Y et al. Molecular cloning of the human cholecystokinin gene by use of a synthetic probe containing deoxyinosine. Proc Natl Acad Sci USA 1985; 82:1931-35.
40. Goltermann NR, Rehfeld JF, Røigaard-Petersen H. In vivo biosynthesis of cholecystokinin in rat cerebral cortex. J Biol Chem 1980; 255:6181-6185.
41. Eng J, Shiina Y, Pan Y-CE, Blacher R et al. Pig brain contains cholecystokinin octapeptide and several cholecystokinin desoctapeptides. Proc Natl Acad Sci USA 1983; 80:6381-6385.
42. Stengaard-Pedersen K, Larsson L-I, Fredens K et al. Modulation of cholecystokinin concentrations in the rat hippocampus by chelation of heavy metals. Proc Natl Acad Sci USA 1984; 81:5876-5880.
43. Rehfeld JF, Hansen HF. Characterization of preprocholecystokinin products in the porcine cerebral cortex: Evidence of different processing pathways. J Biol Chem 1986; 261:5832-40.
44. Kelly RB. Pathways of protein secretion in encaryotes. Science 1985; 230:25-32.
45. Schwartz TW. The processing of peptide precursors. In: Okamoto H, ed. Molecular Biology of Islets of Langerhans. Cambridge University Press, London, 1990:153-205.

46. Eipper BA, Park LP, Dickerson IM et al. Structure of a precursor for an enzyme mediating COOH-terminal amidation in peptide biosynthesis. Mol Endocrinol 1987; 1:777-90.
47. Katopodis AG, Ping D, May SW. A novel enzyme from bovine neurointermediate pituitary catalyzes dealkylation of alpha-hydroxyglycine derivatives thereby functioning sequentially with peptidylglycine alpha-amidating monooxygenase in peptide amidation. Biochemistry 1990; 29:6115-20.
48. Murthy ASN, Mains RE, Eipper BA. Purification and characterization of peptidylglycine x-amidating monooxygenase from bovine neurointermediate pituitary. J Biol Chem 1986; 261:1815-1822.
49. Perkins SN, Husten EJ, Eipper BA. The 108-kDA peptidylglycine alpha-amidating monooxygenase precursor contains two separable enzymatic activities involved in peptide amidation. Biochem Biophys Res Comm 1990; 171:926-32.
50. Eberlein GA, Eysselein VE, Davis MT et al. Patterns of prohormone processing: Order revealed by a new procholecystokinin-derived peptide. J Biol Chem 1992; 267:1517-21.
51. Eng J, Du BH, Pan YCE et al. Purification and sequencing of a rat intestinal 22 amino acid C-terminal CCK-fragment. Peptides 1984; 5:1203-1206.
52. Mutt V. Further investigations on intestinal hormonal polypeptides. Clin Endocrinol 1976; 5:175-84.
53. Reeve JR, Eysselein V, Walsh JH et al. New molecular forms of cholecystokinin. J Biol Chem 1986; 261:16392-97.
54. Dockray GJ, Gregory RA, Hutchison JB et al. Isolation, structure and biological activity of two cholecystokinin octapeptides from sheep brain. Nature 1978; 274:711-713.
55. Shively J, Reeve JR, Eysselein V et al. CCK-5: sequence analysis of a small cholecystokinin from canine brain and intestine. Amer J. Physiol 1987; 252:G272-75.
56. Eysselein V, Eberlein G, Ho FJ et al. An amino-terminal fragment of cholecystokinin-58 is present in the gut: Evidence for a similar processing site of procholecystokinin in canine gut and brain. Reg Peptides 1988; 22:205-15.
57. Cantor P, Rehfeld JF. Cholecystokinin in pig plasma: release of components devoid of a bioactive COOH-terminus. Amer J Physiol 1989; 256:G53-61.
58. Rehfeld JF. The molecular nature of cholecystokinin in plasma - an in vivo immunosorption study in rabbits. Scand J Gastroent 1994; 29:110-21.
59. Yoo OJ, Powell CT, Agarwal K. Molecular cloning and nucleotide sequence of full-length cDNA coding for porcine gastrin. Proc Natl Acad Sci USA 1982; 79:1049-1053.
60. Boel E, Vuust J, Norris K et al. Molecular cloning of human gastrin cDNA: evidence for evolution of gastrin by gene duplication. Proc Natl Acad Sci USA 1983; 80:2866-2869.

61. Marley PD, Rehfeld JF. Extraction techniques for gastrins and cholecystokinins in the rat central nervous system. J Neurochem 1984; 42:1515-1522.
62. Marley PD, Nagy JI, Emson PC et al. Cholecystokinin in the rat spinal cord: distribution and lack of effect of neonatal capsaicin treatment and rhizotomy. Brain Res 1982; 238:494-498.
63. Eysselein VE, Reeve JR, Shively JE et al. Partial structure of a large canine cholecystokinin (CCK-58): amino acid sequence. Peptides 1982; 3:687-691.
64. Rehfeld JF. Tetrin. In eds. Bloom SR, Polak JM. Gut Hormones. Churchill Livingstone, Edinburgh and London. 1981:240-247.
65. Saito E, Sankaran H, Goldfine ID et al. Cholecystokinin receptors in the brain: characterization and distribution. Science 1980; 208:1155-1156.
66. Steigerwalt RW, Williams JA. Binding specificity of the mouse cerebral cortex receptor for small cholecystokinin peptides. Regul Pept 1984; 8:51-59.
67. Sankaran H, Goldfine ID, Deveney CW et al. Binding of cholecystokinin to high affinity receptors on isolated rat pancreatic acini. J Biol Chem 1980; 255:1849-1853.
68. Sankaran H, Bailey AC, Williams JA. CCK-4 contains the full hormonal information of cholecystokinin in isolated pancreatic acini. Biochem Biophys Commun 1981; 103:1356-1362.
69. Rehfeld JF, Larsson L-I, Goltermann N et al. Neural regulation of pancreatic hormone secretion by the C-terminal tetrapeptide of CCK. Nature 1980; 284:33-38.
70. Rehfeld JF. Neuronal cholecystokinin: One or multiple transmitters? J Neurochem 1985; 44:1-10.
71. Gubler U, Chua AO, Hoffman BJ et al. Cloned cDNA to cholecystokinin mRNA predicts an identical preprocholecystokinin in pig brain and gut. Proc Natl Acad Sci USA 1984; 81:4307-4310.
72. Seva C, Dickinson CJ, Yamada T. Growth-promoting effects of glycine-extended progastrin. Science 1994; 265:410-12.
73. Rehfeld JF, Mogensen NW, Bardram L et al. Expression but failing maturation of procholecystokinin in cerebellum. Brain Res 1992; 576:111-19.
74. Rehfeld JF. Accumulation of non-amidated preprogastrin and preprocholecystokinin products in the porcine anterior pituitary. J Biol Chem 1986; 261:5841-47.
75. Rehfeld JF. Preprocholecystokinin processing in the normal human anterior pituitary. Proc Natl Acad Sci USA 1987; 84:3019-24.
76. Rehfeld JF, Johnsen AH, Ødum L et al. Nonsulfated cholecystokinin in human medullary thyroid carcinomas. J Endocrinol 1990; 124 501-6.
77. Bardram L, Hilsted L, Rehfeld JF. Cholecystokinin, gastrin and their precursors in pheochromocytomas. Acta Endocrinol 1989; 120:479-84.

78. Ghatei MA, Sheppard MN, O'Shaughnessy DJ et al. Regulatory peptides in the mammalian respiratory tract. Endocrinol 1982; 111:1248-54.
79. Persson H, Rehfeld JF, Ericsson A et al. Transient expression of the cholecystokinin gene in male germ cells and accumulation of the peptide in the acrosomal granula. Proc Natl Acad Sci USA 1989; 86:6166-70.
80. Snyder SH, Innis RB. Peptide neurotransmitters. Annu Rev Biochem 1979; 48:755-782.
81. Dockray GJ, Gregory RA, Tracy HJ et al. Transport of cholecystokinin octapeptide-like immunoreactivity towards the gut in afferent vagal fibers in cat and dog. J Physiol 1981; 314:501-511.
82. Rehfeld JF, Lundberg JM. Cholecystokinin in feline vagal and sciatic nerves: concentration, molecular form and transport velocity. Brain Res 1983; 275:341-347.
83. Pinget M, Straus E, Yalow RS. Localization of cholecystokinin-like immunoreactivity in isolated nerve terminals. Proc Natl Acad Sci USA 1978; 75:6324-6326.
84. Emson PC, Lee CM, Rehfeld JF. Cholecystokinin peptides: vesicular localization and calcium dependent release from rat brain in vitro. Life Sci 1980; 26:2157-2162.
85. Dodd PR, Edwardson JA, Dockray GJ. The depolarization-induced release of cholecystokinin C-terminal octapeptide (CCK-8) from rat synaptosomes and brain slices. Regul Pept 1980; 1:17-19.
86. Dodd J, Kelly JS. The actions of cholecystokinin and related peptides on pyramidal neurones of the mammalian hippocampus. Brain Res 1981; 205:337-350.
87. Deschodt-Lanckman M, Bui ND, Noyer M et al. Degradation of cholecystokinin-like peptides by a crude rat brain synaptosomal fraction: a study by high-pressure-liquid chromatography. Regul Pept 1981; 2:15-30.
88. Chiodo A, Bunney BS. Proglumide: selective antagonism of excitatory effects of cholecystokinin in central nervous system. Science 1983; 196:998-1000.
89. Collins S, Walker D, Forsyth P et al. The effects of proglumide on cholecystokinin-, bombesin- and glucagon-induced satiety in the rat. Life Sci 1983; 32:2223-2229.
90. Robberecht P, Deschodt-Lanckman M, Vanderhaeghen J-J. Demonstration of biological activity of brain gastrin-like peptidergic material in the human: its relationship with the COOH-terminal octapeptide of cholecystokinin. Proc Natl Acad Sci USA 1978; 75:524-528.
91. Ryder SW, Eng J, Straus E et al. Alkaline extraction of cholecystokinin-immunoreacitivity from rat brain. Biochem Biophys Res Commun 1980; 94:704-709.
92. Ryder SW, Eng J, Straus E et al. Extraction and immunochemical characterization of cholecystokinin-like peptides from pig and rat brain. Proc Natl Acad Sci USA 1981, 78:3892-3896.

93. Barden N, Merand Y, Rouleau D et al. Regional distribution of somatostatin and cholecystokinin-like immunoreactivity in rat and bovine brain. Peptides 1981; 2:299-302.
94. Beinfeld MC, Meyer DK, Eskay RL et al. The distribution of cholecystokinin-immunoreactivity in the central nervous system of the rat as determined by radioimmunoassay. Brain Res 1981; 212:51-57.
95. Hays EE, Beinfeld MC, Jensen RT et al. Demonstration of a putative receptor site for cholecystokinin in rat brain. Neuropeptides 1980; 1:53-62.
96. Van Dijk A, Gillesen D, Möhler H et al. Autoradiographical localisation of cholecystokinin-receptor binding in rat brain and pancreas in vitro using $^{3}$H-CCK-8 as radioligand. Br J Pharmacol 1981; 74:858P.
97. Praissman M, Martinez PA, Saladino CF et al. Characterization of cholecystokinin binding sites in rat cerebral cortex using a $^{125}$I-CCK-8 probe resistant to degradation. J Neurochem 1983; 40:1406-1413.
98. Rehfeld JF. Immunochemical studies on cholecystokinin. I. Development of sequence-specific radioimmunoassays for porcine triacontatriapeptide cholecystokinin. J Biol Chem 1978; 253:4016-4021.
99. Rehfeld JF., ed. Bloom SR. Problems in the technology of radioimmunoassays for gut hormones in Gut Hormones 1978:112-119.
100. Studler JM, Simon H, Cesselin F et al. Biochemical investigation on the localization of the cholecystokinin octapeptide in dopaminergic neurons originating from the ventral tegmental area of the rat. Neuropeptides 1981; 2:131-139.
101. Beinfeld MC, Meyer DK, Brownstein MJ. Cholecystokinin octapeptides in the rat hypothalamo-neurohypophyseal system. Nature 1980; 288:376-378.
102. Straus E, Ryder SW, Eng J et al. Immunochemical studies relating to cholecystokinin in brain and gut. Recent Prog Horm Res 1981; 37:447-475.
103. Beinfeld MC. Chromatographic characterization of gastrin/cholecystokin peptides in bovine and porcine pituitary. Peptides 1982; 3:531-534.
104. Rehfeld JF. Radioimmunochemical analysis of neuropeptides based on general characteristics for the analyte. Trends Anal Chem 1989; 8:102-6.
105. Rehfeld JF, Hansen HF, Larsson L-I et al. Gastrin and cholecystokinin in pituitary neurons. Proc Natl Acad Sci USA 1984; 81:1902-1905.
106. Uvnäs-Wallensten K, Rehfeld JF, Larsson L-I et al. Heptadecapeptide gastrin in the vagal nerve. Proc Natl Acad Sci USA 1977; 74:5707-5710.
107. Rehfeld JF. Progastrin and its products in the cerebellum. Neuropeptides 1991; 20:239-45.

108. Rehfeld JF, Kruse-Larsen C. Gastrin and cholecystokinin in human cerebrospinal fluid: immunochemical studies on concentrations and molecular heterogeneity. Brain Res. 1978; 155:19-26.
109. Kruse-Larsen C, Rehfeld JF. Gastrin in human cerebrospinal fluid: lack of correlation with serum concentrations. Brain Res 1979; 176:189-191.

# CHAPTER 3

# Animal Studies on CCK and Anxiety

**Jaanus Harro, Eero Vasar, Lars Oreland and Pekka T. Männistö**

## ANXIETY-RELATED BEHAVIORAL PHENOMENA WITH CCK RECEPTOR LIGANDS

In 1979, two reports were published that contained hints to some unidentified behavioral action of intracerebrally administered CCK. Thus, Della-Fera and Baile studied the satiety response to CCK-peptides in sheep.[1] When pentagastrin, the common five amino acid sequence of CCK and gastrin, was injected into the cerebral ventricles instead of CCK-8, sheep exhibited behaviors commonly not associated with satiety, such as vocalizations and foot stamping. Retrospectively, one could associate these observations with a panic-like state in these animals. Ishibashi and colleagues performed electrophysiological recordings and mentioned parenthetically that intracerebral injections of CCK-peptides caused agitation in rats.[2] However, no suggestions as to the possibility that CCK-induced neuronal excitation could consistently be associated with emotional states were made. Fekete and coworkers were the first to propose explicitly the anxiogenic potential of neuronal CCK on the basis of avoidance experiments.[3] However, no systematic studies were published until the early 1990s. Recent studies have shown that CCK-peptides administered systemically

*Cholecystokinin and Anxiety: from Neuron to Behavior,* edited by Jacques Bradwejn and Eero Vasar. 

or intracerebrally indeed produce anxiogenic-like effects in several species, including mice, rats, guinea-pigs, cats and monkeys.[4-10]

Manipulations on the CCK receptors can influence anxiety levels in different paradigms. Thus, CCK receptor agonists inhibit exploratory behavior of mice and rats in the elevated plus-maze test, and decrease the time spent and locomotor activity in the light compartment of the light/dark compartment test. CCK-peptides also increase defensive burying in rats.[4] CCK-4 treatment has been shown to suppress feeding in a novel situation and to increase the number of distress calls in the ultrasound vocalization test, in which rat pups are separated from the mother.[11] CCK-peptides support acquisition and retention in fear-motivated memory tests. Even though CCK has been implicated in memory processes not related to anxiety, its effects on appetitively motivated learning in handling-habituated rats are not obvious.[12] Peripherally injected CCK can produce conditioned place aversion in food-deprived rats.[13] The nonselective CCK receptor antagonist proglumide and the selective CCK-B receptor antagonists L-365,260, CI-988 and LY 262691 show anxiolytic-like effects in several animal anxiety tests. Even the selective CCK-A receptor antagonists lorglumide and devazepide show similar properties, but at doses probably active also at CCK-B receptors.[6,14,15] Nevertheless, few published papers have indicated cases in which the anxiolytic effect of CCK-B antagonists as a single treatment has not been evident,[15,16] and there have recently been many reports at scientific meetings that CCK-B antagonists may fail to show anxiolytic-like profiles.

This is theoretically conceivable: since peptide neurotransmission is believed to be activated by bursting or high-frequency neuronal activity, peptide antagonists should not necessarily show any effect under normal tonic activity.[17] Indeed, even though relatively low doses of CCK can be used in drug discrimination training,[18] L-365,260 does not produce readily discriminable interoceptive stimuli in rats.[19] Thus, the anxiolytic effect of a peptide antagonist may be observable only if the neurobiological mechanisms underlying anxiety already are activated.

It is important to stress that the effect of a previous administration of established anxiogenic drugs, both CCK agonists and GABA-negative drugs, can also be blocked by CCK receptor antagonists.[6,8,14,15,20,21] The latter interaction, however, has not been shown consistently.[15]

Some controversy still exists with regard to the CCK receptor subtype involved in the anxiogenic- and anxiolytic-like effects of CCK receptor agonists and antagonists. Thus, Hendrie et al and Bickerdike et al have recently concluded that CCK-A receptor might be involved in the mediation of "generalized anxiety" since devazepide had more reliable anxiolytic-like effects on exploratory behavior than two CCK-B receptor antagonists.[22,23]

One should be cautious with interpretations when the efficacy of CCK receptor antagonists is measured using routine "anxiety" tests. As mentioned above, some investigators have not observed any effect of CCK receptor antagonists on exploratory behavior. On the other hand, the most pronounced "anxiolytic" effects of CCK-B receptor antagonists have been demonstrated using tests based on exploratory activity, frequently exploited to measure anxiety in rodents. In such studies, it is assumed that the test situation brings about some anxiety that can be pharmacologically reduced or enhanced. It is frequently not considered, however, that exploratory behavior is not a simple measure of anxiety. It is rather an evolutionally shaped adaptive strategy, and the presence of neophobia should not be indicative of pathological anxiety.[24] Indeed, exploratory behavior is dependent upon the balance (or interference) between neophobia and the exploratory drive, the latter depending upon multiple intrinsic and environmental factors. The potent anxiolytic effects of CCK receptor antagonists per se (i.e., without previously defined anxiogenic challenge) have been demonstrated using tests of exploratory activity, but not in other tests.[25,26] Thus, CCK-B antagonists are almost inactive in classic "anxiety" paradigms in which conflict is created by delivering direct punishments, e.g., by electric foot-shock. In the four-plate test, CI-988 did not exhibit the usual robust dose-dependent anxiolytic-like profile.[27] Since the four-plate test measures exploration which is punished with foot-shocks, it appears that CCK-B antagonists promote exploration but do not alleviate fear of punishment.

This finding is consistent with the outcome of a meta-analysis of studies on the effect of CCK-B antagonists on exploratory activity: the anxiolytic-like effect is observable when the baseline activity is relatively high (e.g., in plus-maze, more than 20% entries into open arms). We have suggested previously that this remarkable effect of CCK-B receptor antagonists is due to additional effect on the motivation systems that mediate curiosity.[15,28] Indeed, in a choice task, LY-288,513 treatment did not increase exploration of novel areas and objects during the initial (avoidance) period in rats, but did so in rats habituated to the test.[28]

Colocalization of CCK and dopamine in the ventral tegmental area and the ascending mesolimbic pathways suggests that CCK could act as a neuromodulator of dopaminergic transmission.[29,30] These dopaminergic pathways have been closely related to motivational mechanisms and reward and, thus, CCK would have a place in the regulation of motivated behaviours.[31] As could be expected considering the existence of multiple CCKergic subsystems in the mesolimbic area and the complex nature of CCK-dopamine interactions, the actions of CCK receptor agonists and antagonists in the paradigms of motivated behavior are ambiguous.[32] However, microinjection of CCK into the ventral tegmental area potentiates amphetamine-conditioned place preference.[33] Intra-accumbens injections of CCK can either enhance or reduce the behavioral effects of amphetamine and dopamine release dependent upon the injection site and receptor subtype involved, and devazepide can reduce the ability of a rat to discriminate cocaine from saline.[34-36]

Even some of the anxiogenic-like effects of CCK seem to be mediated via these mechanisms. For example, CCK injected into the posterior part of nucleus accumbens reduces novelty-related activity through CCK-A receptors, an effect probably related to reduction of dopamine metabolism by modulation of presynaptic dopamine $D_2$ receptors.[37] Peripherally injected CCK receptor antagonists have therefore the potential to act on distinct GABA- and dopaminergic neural systems that channel their activity into exploratory behavior. Furthermore, since the intensity of CCK-8 effects, at least on the extracellular dopamine level in the posterior nucleus accumbens, depends upon the environmental conditions and the activity of the dopamine neurons, there are even more sources for variable results.[38] Neither have the anxiogenic-like effects of CCK-peptides in animal experiments been observed by all investigators, and this considerable body of negative findings should not be ignored. Even though negative findings often remain unpublished, and the authors have become aware of those by personal communications, few researchers have documented negative findings with CCK-8 or even CCK-4 and CCK-5.[11] Thus, CCK-4 has failed to be aversive in an operant responding paradigm in rats and pentagastrin did not induce anxiety in monkeys.[39-40] Importantly, the effective dose of agonist and behavioral patterns after CCK challenge depend on baseline anxiety of the animal and on the hierarchical position of the animal in its social group. Thus, 10-50 times lower doses of caerulein and pentagastrin are anxiogenic in rats and mice housed under overcrowded conditions compared to animals housed under noncrowded conditions, whereas social isolation further decreases the sensitivity of rats to caerulein challenge.[20,41,42] In monkeys, the "uptight" animals, typically restless, submissive to threat, and excessively reactive to environment, become anxious after low doses of CCK-4, and the behavior of basically calm conspecifics seems to be rather different after CCK-4 injection.[8]

These apparently contradictory findings suggest that CCK-B receptor stimulation elicits anxiety only if the animal is in distress. As to neurochemistry, CCK-4 treatment potentiates the increase in extracellular 5-HT levels in the lateral prefrontal cortex induced by the exposure of guinea-pigs to the elevated plus-maze, but has no effect when the animals remain in their home-cage throughout the experiment.[10]

## SITE OF THE ANXIOGENIC ACTION OF CCK IN RODENTS

The site of the anxiogenic action of CCK agonists has remained unknown. Several behavioral effects of peripherally administered CCK-peptides are mediated through the nervus vagus.[43] However, cerulein treatment was able to reduce plus-maze activity in rats after subdiaphragmatic vagotomy.[44] Currently, there is no available evidence that CCK-4 crosses the blood-brain barrier. However, the possibility exists that CCK-4 affects CCK-B receptors in brain regions which are not fully protected by the blood-brain barrier. We have proposed that the brainstem regions that ordinarily function to monitor sympathetic nerve discharge and vasomotor tone are involved in the panicogenic effect of peripherally administered CCK-peptides.[45] Further, there are experimental data suggesting that CCK interacts with these brainstem mechanisms via both CCK-A and CCK-B receptors.[46] Peripheral administration of CCK leads to increased c-fos expression and thus increased cellular activity of the nucleus of the solitary tract and area postrema.[47] Recent finding that intraperitoneal, but not intrahippocampal or intracerebroventricular, administration of

L-365,260 elicits anxiolytic effect further corroborates this hypothesis.[48] Intracerebrally injected CCK-peptides can be anxiogenic at very low doses, when given, e.g., into central amygdala.[3,4,49] Thus, a brainstem site of anxiogenic action of CCK does not exclude additional higher brain sites of action, which could be affected by neuronal CCK. Nonpeptide CCK receptor antagonists could therefore block the CCK-mediated mechanisms leading to anxiety by acting at different sites in the brain. These include the brainstem, the limbic system and the frontal association area.[45] Changes observed in the brain CCK levels and receptor binding in stressful situations seem to support such an assumption.

## CHANGES IN THE CCKERGIC MECHANISMS IN ANXIETY STATES

Few animal and human studies have addressed the question of whether or not the alterations in endogenous CCKergic mechanisms can be associated to anxiety states. An early report pointed to the increased levels of CCK-like immunoreactivity in limbic regions and in the prefrontal cortex in rats in response to unavoidable electric foot-shocks.[50] Alterations in CCK receptor binding after different anxiogenic challenges have been described.[21,42,51-53] Namely, treatment with FG 7142, an anxiogenic β-carboline, or with picrotoxin, the noncompetitive GABA antagonist, or administration of NMDA at an anxiogenic dose, social isolation, or withdrawal from long-term diazepam administration, all cause upregulation of [$^3$H]propionylated-CCK-8 binding in the frontal cortex in the rat. This change may be of physiological importance, since high CCK receptor binding has been observed in rats selected according to an extremely low exploratory activity in the elevated plus-maze. A hypothesis was made that high cortical CCK receptor density is a correlate of anticipatory anxiety and/or deficits in adaptation with the environment.[51] This hypothesis has gained support from a human post-mortem study[54] in which the CCK receptor binding in the frontal and cingulate cortices of suicide victims was found to be higher compared to well-matched control cases. One might speculate that a high density of CCK receptors in cortical regions is a basic biological correlate of deficits in adaptation with environment.

Recently, preproCCK mRNA levels were measured after acute and chronic diazepam treatment and diazepam withdrawal.[55] CCK mRNA levels were elevated in the cerebral cortex and in hippocampal non-pyramidal neurons after acute diazepam administration as well as after withdrawal from 3-week diazepam treatment, whereas no change was observable 30 min after last injection of diazepam after 3 week treatment period. Diazepam withdrawal, but not acute diazepam administration produces upregulation of CCK receptor binding.[52] Thus, it seems that while increased expression and release of CCK-peptide(s) is a correlate of environmental challenge, it can become a correlate of pathological anxiety under the conditions where CCK receptors are upregulated.[50,56] In line with this, CCK-B receptor antagonist LY-288,513 treatment has no effect on auditory startle reflex in rats per se, but blocks the increasing effect of diazepam withdrawal.[57] CCK-B receptor blockade also prevents the gastrointestinal effects of diazepam withdrawal.[58] The above hypothesis would be consistent with the findings of Wiertelak and colleagues that endogenous CCK can mediate, through CCK-B receptors, the signals of safe environment.[59] Indeed, a recent study has demonstrated that subchronic treatment with antipanic drugs increases CCK levels in rat ventral tegmental area and limbic regions.[60] CCK levels and receptor binding have been investigated in parallel in one study.[61] A stressful event (housing of rats in the close neighborhood of a place where other conspecifics were being sacrificed) led to the upregulation of CCK receptor binding in the frontal cortex and to an increase in the CCK levels in the hippocampus.

In conclusion, both CCK levels and CCK receptor characteristics undergo changes in states of anxiety and/or adaptation deficit, but the effect depends on the brain region and the nature of challenging factors. An alternative CCK-related mechanism that might operate in anxious states is an alteration in peptide processing. The existence of distinct post-translational processing pathways in the brain has been proposed.[62]

Since selective CCK-B receptor activation leads to increased anxiety, it is tempting to speculate that a selective increase in the production of some short fragments of CCK could be responsible. Indeed, recently it was demonstrated in an ethologically relevant paradigm (cat odor exposure), that anxious/helpless pattern of behavior correlates with increased levels of CCK-4 in many brain regions.[63] The concentration of CCK-8 was increased only in the ventral striatum. To which extent differential peptide processing, increased CCK-8 levels, and enhanced receptor sensitivity contribute to the development of different anxious states remains to be elucidated.

## ANIMAL STUDIES ON ANXIETY-RELATED NEUROTRANSMITTER INTERACTIONS

Early studies have indicated a possible relationship between the anxiogenic effect of CCK-peptides and the GABAergic mechanisms. Bradwejn and de Montigny demonstrated that benzodiazepine receptor agonists (flurazepam, chlordiazepoxide, lorazepam or diazepam) selectively and specifically antagonized CCK-8S-induced excitation of hippocampal pyramidal neurons in rats, and that this effect was reversed by pretreatment with the benzodiazepine receptor antagonist flumazenil.[64] Additional experiments with PK 8165, a partial benzodiazepine receptor agonist with selective anxiolytic effect, revealed that the antagonism by benzodiazepines of CCK-8S-induced excitation was specifically related to their anxiolytic action.[65] ED50 values of lorazepam and diazepam (32 ug/kg and 106 ug/kg, respectively) on excitations produced by CCK-8S fall within the range of clinically used doses for these medications. These findings seem to be compatible with the evidence that CCKergic and GABAergic neurons are interactive. Thus, the release of CCK is under tonic control of GABA in rat cerebral cortex and, in the cerebral cortex, hippocampus and basolateral amygdala, CCK is localized in a fraction of GABA-ergic neurons.[66-69]

However, the functional significance of such a colocalization is not clear for the time being. Changes in CCK mRNA and CCK receptor binding have been reported in benzodiazepine withdrawal (see above), and could contribute to the rebound anxiety after discontinuation of benzodiazepine administration. Animal studies have shown that CCK-4 treatment can potentiate the release of 5-HT in the guinea pig frontal cortex induced by exposure to the elevated plus-maze and increase the foot-shock induced defensive aggressiveness in male rats.[10,70] On the other hand, 5-HT can increase the release of CCK in the rat cerebral cortex and nucleus accumbens.[71] The effect of CCK-4 on defensive aggressiveness was blocked by pretreatment with pirenperone, a 5-$HT_2$ receptor antagonist, whereas the anxiogenic-like effect of caerulein on mice exploratory activity was attenuated by 5-$HT_3$ receptor antagonist ondansetron at one dose level.[41,70] Further studies are required to clarify the possibly complex role of 5-HT-ergic neurons in the anxiogenic-like actions of CCK-peptides.

Less information is available concerning the possible noradrenergic mechanisms linked to the effects elicited by CCK-peptides. Monstein et al have shown that noradrenaline (NA) can enhance the expression of preproCCK.[72] The effect of DSP-4, a neurotoxin preferentially damaging the axon terminals of the projections from the locus ceruleus, on CCK receptor characteristics has been described.[73,74] Treatment of rats with DSP-4 does not cause profound behavioral changes, but seems to impair rather selectively their response to novelty.[75] Archer has hypothesized, on the basis of his studies on avoidance learning, that DSP-4 treated rats fail to engage in behavior necessary to cope with environmental changes.[76] Our recent behavioral and pharmacological analysis supports this claim and suggests that the decline in neotic behavior after DSP-4 is primarily due to an increase in neophobia, even though a decrease in curiosity could not be ruled out.[77] DSP-4 administration causes significant dose-dependent upregulation of CCK receptor binding in the frontal cortex and hippocampus in rat brain, an effect that can be prevented by a desipramine pretreatment before DSP-4 injection. Comparing the time curves of the effect of DSP-4 on NA uptake and CCK receptor binding, it seems that the upregulation of CCK receptors follows to the degeneration of NA-containing nerve terminals.[74] This finding, in line with the CCK receptor changes in anxiety states (see above), suggests that deficiency in NA-ergic neurotransmission can lead to alterations in CCKergic mechanisms possibly related to anxiety.

Several investigators have recently proposed a role of glutamatergic neurotransmission in anxiety. Glutamate receptor agonists enhance the release of CCK in the rat cerebral cortex.[66] Low doses of N-methyl-D-aspartate (NMDA) have been shown to have anxiogenic-like effect on mouse exploratory behavior.[53] Proglumide, an unselective CCK receptor agonist, attenuated this effect of NMDA treatment, but this was not the case with devazepide and L-365,260 treatment. Devazepide, at low doses, only tended to reduce the effect of NMDA, whereas the treatment with L-365,260 actually potentiated the effect of NMDA. Obviously, both CCK receptor subtypes should be blocked to reduce the effect of NMDA. One explanation to this curious finding is that L-365,260 treatment blocks the normal CCK-B receptor-mediated control of the glutamatergic neurotransmission,[77,78] and thus leads to hypersensitivity to NMDA.

An interesting aspect of the multiple effects of CCK is its influence on the hypothalamic-pituitary-adrenal function. Both peripheral and intracerebral administration of CCK-8 increase corticotropin releasing factor (CRF) release and levels of ACTH and corticosterone in plasma.[79] Peripheral administration of CCK activates c-fos expression in hypothalamic CRF-secreting neurons.[80] The anxiogenic-like effect of CCK-8 on the rat exploratory behavior can be blocked by pretreatment with CRF antiserum or CRF receptor antagonist.[9] ACTH and cortisol release are facilitated by CCK in humans, but a study which examined the ACTH response to pentagastrin in normal subjects and panic disorder patients found that those subjects having the most intense symptoms of panic had the smallest ACTH responses.[81] These findings suggest that the CRF-ergic mechanisms get activated by CCK in acute stress situations. Subsequently, high levels of CRF reduce the activity of the CCKergic mechanisms as an adaptational response. In line with this, Takamatsu and colleagues have shown that CRF elicits a profound reduction of the CCK concentration in the rat medial frontal and anterior cingulate cortex.[82] In conclusion, a hypothesis can be made that CCK is a normal endogenous anticipatory stress modulator that could be involved in pathological anxiety due to some alteration(s) in the normal regulation of CCKergic mechanisms.

## REFERENCES

1. Della-Fera MA, Baile CA. Cholecystokinin octapeptide: Continuous picomole injections into the cerebral ventricles of sheep suppress feeding. Science 1979; 206:471-473.
2. Ishibashi S, Oomura Y, Okajima T et al. Cholecystokinin, motilin and secretin effects on the central nervous system. Physiol Behav 1979; 23:401-403.
3. Fekete M, Lengyel A, Hegedues B et al. Further analysis of the effects of cholecystokinin octapeptide on avoidance behavior in rats. Eur J Pharmacol. 1984; 98:79-91.
4. Csonka E, Fekete M, Nagy G et al. Anxiogenic effect of cholecystokinin in rats. In: eds Penke B, Török A. Peptides 1988:249-252. Walter de Gruyter & Co, New York.
5. Harro J, Pold M, Vasar E. Anxiogenic-like action of caerulein, a CCK-8 receptor agonist, in the mouse: influence of acute and subchronic diazepam treatment. Naunyn-Schmiedeberg's Arch Pharmacol 1990; 341:62-67.

6. Harro J, Vasar E. Evidence that CCK-B receptors mediate the regulation of exploratory behaviour in the rat. Eur J Pharmacol 1991; 193:379-381.
7. Singh L, Lewis AS, Field MJ et al. Evidence for an involvement of the brain cholecystokinin B receptor in anxiety. Proc Natl Acad Sci USA 1991; 88:1130-1133.
8. Palmour RM, Bradwejn J, Ervin FR. The anxiogenic effects of CCK-4 in monkeys are reduced by CCK-B antagonists, benzodiazepines or adenosine A2 agonists. Eur Neuropsychopharmacol 1992; 2:193-195.
9. Biro E, Sarnyai Z, Penke B et al. Role of endogenous corticotropin-releasing factor in mediation of neuroendocrine and behavioral responses to cholecystokinin octapeptide sulfate ester in rats. Neuroendocrinology1993; 57:340-345.
10. Rex A, Fink H, Marsden CA. Effects of Boc-CCK-4 and L 365,260 on cortical 5-HT release in guinea-pigs on exposure to the elevated plus maze. Neuropharmacology 1994; 33:559-565.
11. Rex A, Barth T, Voigt J-P et al. Effects of cholecystokinin tetrapeptide and sulfated cholecystokinin octapeptide in rat models of anxiety. Neurosci Lett 1994; 172:139-142.
12. Harro J, Oreland L. Cholecystokinin receptors and memory: a radial maze study. Pharmacol Biochem Behav 1993; 44:509-519.
13. Swerdlow NR, van der Kooy D, Koob GF et al. Cholecystokinin produces conditioned place-aversions, not place-preferences, in food-deprived rats: Evidence against involvement in satiety. Life Sci 1983; 32:2087-2093.
14. Woodruff GN, Hughes J. Cholecystokinin antagonists. Annu Rev Pharmacol Toxicol 1991; 31:469-501.
15. Harro J, Vasar E. Cholecystokinin-induced anxiety: How is it reflected in studies on exploratory behaviour? Neurosci Biobehav Rev 1991; 15:473-477.
16. Crawley JN. Subtype-selective cholecystokinin receptor antagonists block cholecystokinin modulation of dopamine-mediated behaviors in the rat mesolimbic pathway. J Neurosci 1992; 12:3380-3391.
17. Hökfelt T. Neuropeptides in perspective: The last ten years. Neuron 1991; 7:867-879.
18. Melton PM, Kopman JA, Riley AL. Cholecystokinin as a stimulus in drug discrimination learning. Pharmacol Biochem Behav 1993; 44:249-252.
19. Jackson A, Tattersall D, Bentley G et al. An investigation into the discriminative stimulus and reinforcing properties of the CCK-B-receptor antagonist, L-365,260 in rats. Neuropeptides 1994; 26:343-353.
20. Harro J, Pold M, Vasar E et al. The role of CCK-8-ergic mechanisms in the regulation of emotional behaviour in rodents. J Higher Nerv Act 1989; 39:877-883.

21. Vasar E, Harro J, Pold A et al. CCK receptors and anxiety in rats. In: eds. Dourish CT, Cooper SJ, Iversen SD, Iversen LL. Multiple Cholecystokinin Receptors in the CNS, 1992:143-148. Oxford University Press, New York.
22. Hendrie CA, Neill JC, Shepherd JK et al. The effects of CCK-A and CCK-B antagonists on activity in the black/white exploration model of anxiety in mice. Physiol Behav 1993; 54:689.
23. Bickerdike MJ, Marsden CA, Dourish CT et al. The influence of 5-hydroxytryptamine re-uptake blockade on CCK receptor antagonist effects in the rat elevated zero-maze. Eur J Pharmacol 1994; 271:403-411.
24. Harro J. Measurement of exploratory behaviour in rodents. In: ed Conn PM. Methods in Neurosciences, 1993; 14:359-377. Academic Press, San Diego.
25. Hughes J, Boden P, Costall B et al. Development of a class of selective cholecystokinin type B receptor antagonists having potent anxiolytic activity. Proc Natl Acad Sci USA 1990; 87: 6728-6732.
26. Powell KR, Barrett JE. Evaluation of the effects of PD 134308 (CI-988), a CCK-B antagonist, on the punished responding of squirrel monkeys. Neuropeptides 1991; 19(Suppl):75-78.
27. Dooley DJ, Klamt I. Differential profile of the CCK-B receptor antagonist CI-988 and diazepam in the four-plate test. Psychopharmacol 1993; 112:452-454.
28. Harro J. Studies on the Brain Cholecystokinin Receptors and Behaviour. Acta Universitatis Upsaliensis 1993; 421:48 Uppsala University Press.
29. Hökfelt T, Rehfeld JF, Skirboll L et al. Evidence for co-existence for dopamine and CCK in mesolimbic neurons. Nature 1980; 285:476-478.
30. Vanderhaeghen JJ, Lotstra F, Demey J et al. Immunohistochemical localization of cholecystokinin- and gastrin-like peptides in the brain and hypophysis of the rat. Proc Natl Acad Sci USA 1980; 77:1190-1194.
31. Wise RA, Rompre PP. Brain dopamine and reward. Annu Rev Psychol 1989; 40:191-225.
32. Crawley JN. Cholecystokinin-dopamine interactions. Trends Pharmacol Sci 1991; 12:232-236.
33. Pettit HO, Mueller K. Infusions of cholecystokinin octapeptide into the ventral tegmental area potentiate amphetamine conditioned place preferences. Psychopharmacol 1989; 99:423-426.
34. Vaccarino FJ, Rankin J. Nucleus accumbens cholecystokinin (CCK) can either attenuate or potentiate amphetamine-induced locomotor activity: Evidence for rostral-caudal differences in accumbens CCK function. Behav Neurosci 1989; 103:831-836.
35. Marshall FH, Barnes S, Hughes J et al. Cholecystokinin modulates the release of dopamine from the anterior and posterior nucleus accumbens by two different mechanisms. J Neurochem 1991; 56:917-922.

36. Massey BW, Vanover KE, Woolverton WL. Effects of cholecystokinin antagonists on the discriminative stimulus effects of cocaine in rats and monkeys. Drug Alcohol Dependence 1994; 34, 105-111.
37. Derrien M, Durieux C, Daugé V et al. Involvement of $D_2$ dopaminergic receptors in the emotional and motivational responses induced by injection of CCK-8 in the posterior part of the rat nucleus accumbens. Brain Res 1993; 617:181-188.
38. Ladurelle N, Daugé V, Roques BP. Influence of environmental conditions on CCK-8-induced dopamine release in the posterior nucleus accumbens: a microdialysis study in freely moving rats. Neuropeptides 1993; 26S1:23.
39. Bayley PJ, Dawson GR. The effect of i.v. administration of CCK-4 on lever pressing rates of rats on an operant random interval schedule. Br J Pharmacol 1993; 108S:244P.
40. Rupniak NMJ, Schaffer L, Siegel P et al. Failure of intravenous pentagastrin challenge to induce panic-like effects in rhesus monkeys. Neuropeptides 1993; 25:115-119.
41. Vasar E, Peuranen E, Ööpik T et al. Ondansetron, an antagonist of 5-$HT_3$ receptors, antagonizes the anti-exploratory effect of caerulein, an agonist of CCK receptors, in the elevated plus-maze. Psychopharmacology 1993; 110:213-218.
42. Vasar E, Peuranen E, Harro J et al. Social isolation of rats increases the density of cholecystokinin receptors in the frontal cortex and abolishes the anti-exploratory effect of caerulein. Naunyn Schmiedebergs Arch Pharmacol 1993; 348:96-101.
43. Morley JE. Behavioral effects of administered cholecystokinin. ISI Atlas Pharmacol 1987; 1:49-51.
44. Vasar E, Lang A, Harro J et al. Subdiaphragmatic vagotomy does not prevent the anti-exploratory effect of caerulein in the elevated plus-maze. Neuropeptides 1994; 26:39-45.
45. Harro J, Vasar E, Bradwejn J. CCK in animal and human research on anxiety. Trends Pharmacol Sci 1993; 14:244-249.
46. Branchereau P, Böhme GA, Champagnat J et al. Cholecystokinin-A and cholecystokinin-B receptors in neurons of the brainstem solitary complex of the rat: Pharmacological identification. J Pharmacol Exp Ther 1992; 260:1433-1440.
47. Rinaman L, Verbalis JG, Sricker EM et al. Distribution and neurochemical phenotypes of caudal medullary neurons activated to express cFos following peripheral administration of cholecystokinin. J Comp Neurol 1993; 338:475-490.
48. Sondern U, Voits M, Marsden CA et al. Is the anxiolytic effect of the CCK-B receptor antagonist L 365,260 mediated by a central or a peripheral action? Naunyn-Schmiedeberg's Arch Pharmacol 1994; 349S:R89.
49. Belcheva I, Belcheva S, Petkov VV et al. Asymmetry in behavioral responses to cholecystokinin microinjected into rat nucleus accumbens and amygdala. Neuropharmacology 1994; 33:995-1002.

50. Siegel RA, Dueker E-M, Fuchs E et al. Responsiveness of mesolimbic, mesocortical, septal and hippocampal cholecystokinin and substance P neuronal systems to stress, in the male rat. Neurochem. Int 1995; 6:783-789.
51. Harro J, Kiivet R-A, Lang A et al. Rats with anxious or non-anxious type of exploratory behaviour differ in their brain CCK-8 and benzodiazepine receptor characteristics. Behav Brain Res 1990; 39:63-71.
52. Harro J, Lang A, Vasar E. Long-term diazepam treatment produces changes in cholecystokinin receptor binding in rat brain. Eur J Pharmacol 1990; 180:77-83.
53. Vasar E, Harro J, Lang A et al. Anti-exploratory effect of N-methyl-D-aspartate in elevated plus-maze. Involvement of NMDA and CCK receptors. Eur Neuropsychopharmacol 1993; 3:63-73.
54. Harro J, Marcusson J, Oreland L. Alterations in brain cholecystokinin receptors in suicide victims. Eur Neuropsychopharmacol 1992; 2:57-63.
55. Rattray M, Singhvi S, Wu, P-I et al. Benzodiazepines increase preprocholecystokinin messenger RNA levels in rat brain. Eur J Pharmacol [Mol Pharmacol Sect].1993; 245:193-196.
56. Rosen A, Brodin K, Eneroth P et al. Short-term restraint stress and s.c. saline injection alter the tissue levels of substance P and cholecystokinin in the peri-aquaeductal grey and limbic regions of rat brain. Acta Physiol Scand 1992; 146:341-348.
57. Rasmussen K, Helton DR, Berger JE et al. The CCK-B antagonist LY288513 blocks effects of diazepam withdrawal on auditory startle. NeuroReport 1993; 5:154-156.
58. Bonnafous C, Martinez J, Bueno L. Gastrointestinal effects of diazepam-withdrawal are linked to activation of central cholecystokinin-ergic pathways in rats. J Pharm Pharmacol 1994; 46:784-788.
59. Wiertelak EP, Maier SF, Watkins LR. Cholecystokinin antianalgesia: Safety cues abolish morphine analgesia. Science 1992; 256:830-833.
60. Brodin K, Ögren S-O, Brodin E. Clomipramine and clonazepam increase cholecystokinin levels in rat ventral tegmental area and limbic regions. Eur J Pharmacol 1994; 263:175-180.
61. Harro J, Löfberg C, Rehfeld J et al. Brain cholecystokinin levels and receptor binding in anxiety states. Eur Neuropsychopharmacol 1994; 4:352.
62. Rehfeld JF, Hansen HF. Characterization of preprocholecystokinin products in the porcine cerebral cortex: evidence of different processing pathways. J Biol Chem 1986; 261, 5832-5840.
63. Pavlasevic S, Bednar I, Qureshi GA et al. Brain cholecystokinin tetrapeptide levels are increased in a rat model of anxiety. Neuroreport 1993; 5:225-228.
64. Bradwejn J, de Montigny C. Benzodiazepines antagonize cholecystokinin-induced activation of rat hippocampal neurones. Nature 1984; 312:363-364.

65. Bradwejn J, de Montigny C. Effects of PK 8165, a partial benzodiazepine receptor agonist, on cholecystokinin-induced activation of hippocampal pyramidal neurons: a microiontophoretic study in the rat. Eur J Pharmacol 1985; 112:415-418.
66. Yaksh TL, Furui T, Kanawati IS et al. Release of cholecystokinin from rat cerebral cortex in vivo: role of GABA and glutamate receptor systems. Brain Res 1987; 406:207-214.
67. Hendry SHC, Jones EG, De Felipe J et al. Neuropeptide-containing neurons of the cerebral cortex are also GABAergic. Proc Natl Acad Sci USA 1984; 81:6526-6530.
68. Somogyi P, Hodgson AJ, Smith AD et al. Different populations of GABAergic neurons in the visual cortex and hippocampus of cat contain somatostatin- or cholecystokinin-immunoreactive material. J Neurosci 1984; 4:2590-2603.
69. McDonald AJ, Pearson JC. Coexistence of GABA and peptide immunoreactivity in non-pyramidal neurons of the basolateral amygdala. Neurosci Lett 1989; 100:53-58.
70. Vasar E, Maimets M, Allikmets L. Role of serotonin$_2$-receptors in the regulation of aggressive behaviour. Zh Vyss Nerv Deyat 1984; 34:283-289.
71. Paudice P, Raiteri M. Cholecystokinin release mediated by 5-HT$_3$ receptors in rat cerebral cortex and nucleus accumbens. Br J Pharmacol 1991; 103:1790-1794.
72. Monstein HJ, Folkesson R, Geijer T. Procholecystokinin and proenkephalin A mRNA expression is modulated by cyclic AMP and noradrenaline. J Mol Endocrinol 1990; 4:37-41.
73. Fritschy J-M, Grzanna R. Immunohistochemical analysis of the neurotoxic effects of DSP-4 identifies two populations of noradrenergic axon terminals. Neuroscience 1989; 30:181-197.
74. Harro J, Jossan SS, Oreland L. Changes in cholecystokinin receptor binding in rat brain after selective damage of locus ceruleus projections by DSP-4 treatment. Naunyn - Schmiedeberg's Arch Pharmacol 1992; 346:425-431.
75. Delini-Stula A, Mogilnicka E, Hunn C et al. Novelty-oriented behavior in the rat after selective damage of locus ceruleus projections by DSP-4, a new noradrenergic neurotoxin. Pharmacol Biochem Behav 1984; 20:613-618.
76. Archer T, Söderberg U, Ross SB et al. Role of olfactory bulbectomy and DSP-4 treatment in avoidance learning in the rat. Behav Neurosci 1984; 98:496-505.
77. Harro J, Westerling P, Oreland L. CCK-B receptor activation reduces glutamate-induced depolarization in slices of rat cerebral cortex. J Neural Transm [Gen Sect] 1993; 93:61-66.
78. Shinohara S, Katsuura G, Eigyo M et al. Inhibitory effect of CCK-8 and ceruletide on glutamate-induced rises in intracellular free calcium concentrations in rat neuron cultures. Brain Res 1992; 588:223-228.

79. Kamilaris TC, Johnson EO, Calogero AE et al. Cholecystokinin-octapeptide stimulates hypothalamic- pituitary-adrenal function in rats: Role of corticotropin-releasing hormone. Endocrinology 1992; 130:1764-1774.
80. Verbalis JG, Stricker EM, Robinson AG et al. Cholecystokinin activates c-Fos expression in hypothalamic oxytocin and corticotropin-releasing hormone neurons. J Neuroendocrinol 1991; 3:205-213.
81. Abelson JL, Nesse RM, Vinik A. Stimulation of corticotropin release by pentagastrin in normal subjects and patients with panic disorder. Biol Psychiat 1991; 29:1220-1223.
82. Takamatsu Y, Yamamoto H, Ogunremi OO, et al. The effects of corticotropin-releasing hormone on peptidergic neurons in the rat forebrain. Neuropeptides 1991; 20:255-265.

# CHAPTER 4

# CHOLECYSTOKININ AND PANIC DISORDER

**Jacques Bradwejn**

## CCK HYPOTHESIS OF PANIC DISORDER: HISTORICAL PERSPECTIVE

The hypothesis that CCK may be a mediator of anxiety originated from electrophysiological experiments of Bradwejn and de Montigny[1] which demonstrated that anxiolytic molecules of the benzodiazepine family (flurazepam, chlordiazepoxide, lorazepam or diazepam) selectively and specifically antagonized CCK-8S-induced excitation of hippocampal pyramidal neurons in rats, and that this effect was reversed by pretreatment with the benzodiazepine receptor antagonist flumazenil. Dose response curves of the intravenous effect of lorazepam or diazepam on excitations produced by CCK-8S revealed ED50 (effective dose for 50% inhibition) values of 32 µg per kilo and 106 µg per kilo, respectively. These values fell within the range of clinically used doses for these medications.

This study provided the first evidence that anxiolytic benzodiazepines could antagonize the central action of a neuropeptide, and it was proposed that benzodiazepine-mediated antagonism of CCK-induced excitation might be an important mechanism by which benzodiazepines exert their clinically relevant action. More importantly, the observation that an anxiolytic could block the excitatory action of CCK raised questions about whether CCK might be an endogenous anxiogen. Two pilot studies, one in pa-

*Cholecystokinin and Anxiety: from Neuron to Behavior,* edited by Jacques Bradwejn and Eero Vasar. 

tients suffering from panic disorder and the other in healthy subjects with no personal or family history of panic attacks, were conducted to address this question using the tetrapeptide form of CCK (CCK-4). The decision to administer the tetrapeptide form to patients with panic disorder was based on anecdotal data presented by Jens Rehfeld at the Neuronal Cholecystokinin Conference in Brussels in 1984. In the course of investigating the neuroendocrine effects of CCK-4 in healthy human subjects, Jens Rehfeld noted that CCK-4 produced a number "side effects" such as anxiety, dyspnea and depersonalization.[2,3] In our assessment, these "side effects" were strikingly similar to symptoms experienced by panic patients during their spontaneous panic attacks.

Bradwejn and colleagues first administered CCK-4 to patients suffering from panic disorder using a double-blind placebo control methodology. Bolus injections of CCK-4 (50 μg) precipitated a panic attack, as defined by DSM-III criteria and patient self-report, within 1 minute following administration in 11 patients studied, whereas none of the patients panicked following placebo.[4] In addition, CCK-4 elicited an average of 12 symptoms per patient, the most common symptoms being dyspnea, palpitations/rapid heart, chest pain/discomfort, faintness, dizziness, paraesthesia, hot flushes/cold chills, nausea/abdominal distress, anxiety/fear/apprehension and fear of losing control. De Montigny first reported that exogenous CCK-4 produced "panic-like" attacks in healthy volunteers and that these effects could be attenuated by pretreatment with lorazepam.[5]

Taken together, these preliminary data suggested a potential link between CCK activity and panic disorder.

## VALIDATION OF CCK-4 AS A PANICOGENIC AGENT

Our finding of a close analogy between symptoms produced by CCK-4 and those reported to occur during patients' spontaneous panic attacks was intriguing and suggested that CCK-4 might be a suitable paradigm for studying the neurobiology of panic disorder and for anxiety research in general. An important task we set for ourselves was to systematically evaluate the validity of CCK-4 as a model of panic using the seven criteria for an "ideal" panicogenic agent described by Guttmacher[6] and Gorman et al.[7] The seven criteria are as follows:

### THE AGENT SHOULD BE SAFE

CCK-4 seems safe to administer to human subjects. Our experiments done with over 300 subjects together with experiments of other researchers suggest that the peptide is safe for human research. With the exception of a brief vaso-vagal reaction occurring in less than 3% of subjects, no significant adverse effects have been observed.

### THE AGENT SHOULD INDUCE AFFECTIVE AS WELL AS SOMATIC SYMPTOMS OF A PANIC ATTACK

CCK-4 generates both emotional (e.g., anxiety, fear, apprehension) and somatic symptoms (e.g., dyspnea, palpitations, choking, sweating, faintness) that typically occur during a panic attack.[8] In our studies, the occurrence of a subjective sense of anxiety, fear and/or apprehension, as well as at least four DSM-III-R somatic symptoms, are important criteria for judging the occurrence of a panic attack.

### THE AGENT SHOULD PROVOKE ATTACKS WHICH RESEMBLE THE PATIENT'S CLINICAL PANIC ATTACKS

The panic attacks induced by CCK-4 have been appraised by patients to be identical or very similar to their spontaneous panic attacks in terms of the type and quality of symptoms.[9] This has been an important criterion of panic attack in our studies with patients. Moreover, CCK-4 does not induce a stereotyped response in patients. Rather it mimics the individual symptom profile usually experienced by each patient. The majority of patients have reported that the main difference between the CCK-4-induced panic attack and their clinical attacks is that the symptoms induced with CCK-4 occur more abruptly and are generally of a shorter duration.

### THE EFFECTS OF THE AGENT SHOULD BE SPECIFIC FOR PATIENTS WITH A HISTORY OF PANIC ATTACKS

We have found that response to CCK-4 reliably differentiates panic disorder patients from healthy controls with no personal or family history of panic attacks. In a double-blind placebo control study we noted that patients with panic disorder experienced a greater number of symptoms and more intense symptoms following challenge with two doses of CCK-4 (25 and 50 μg).[9]

In addition, the incidence of panic attacks was markedly higher in patients than controls following injection of 25 μg (91% versus 17%) and 50 μg (100% versus 47%) of the peptide. Interestingly, we noted that the number and intensity of symptoms as well as the symptom profile were remarkably similar in both patients and normal subjects who panicked with the 50 μg dose of CCK-4, suggesting that the enhanced response to CCK in patients could not be readily attributed to a tendency to overendorse symptoms. Our results are corroborated by studies by Abelson and Nesse using pentagastrin, a CCK agonist which incorporates the identical 4-amino acid sequence of CCK-4 and by van Megen and colleagues using CCK-4.[10,11,12] These authors found that pentagastrin or CCK-4 provoked panic attacks at a higher frequency in panic disorder patients than in healthy subjects. Further studies are required to determine whether the effects of CCK-4 can differentiate patients suffering from panic disorder and those with other psychiatric syndromes.

### THE EFFECTS OF THE AGENT SHOULD BE RELIABLE

In order to determine whether the behavioral effects of CCK-4 could be replicated in the same individual we administered 25 μg of CCK-4 to 11 panic patients on two separate occasions in the absence of intervening treatment.[13] Although the latency to effect symptoms with CCK-4 was significantly shorter on the second challenge day, the vulnerability of patients to the panicogenic properties of CCK-4 was undiminished with repeated challenge. Panic attack frequency following the initial and subsequent challenge was 82% and 73%, respectively. In addition, the number and intensity of symptoms remained constant with rechallenge.

The effectiveness of CCK-4 in provoking panic responses also appears to be dose dependent. In a double-blind dose response study of CCK-4 (0, 10, 15, 20 and 25 μg) in patients with panic disorder, a significant linear relationship was found for the number and sum intensity of symptoms evoked with CCK-4.[14] All doses of CCK-4 produced a significantly greater number of symptoms than placebo, whereas the sum intensity scores were significantly higher with the 15, 20 and 25 μg dose of CCK-4 than with placebo. The panic attack rate was 17% (10 μg), 64% (15 μg), 75% (20 μg) and 75% (25 μg). None of the patients panicked with placebo (0 μg). The difference between treatments in panic frequency was significantly different and there was a significant linear dose-response effect. Paralleling the behavioral changes induced with CCK-4, a marked and dose-related increase in heart rate and blood pressure was evident. In another double-blind study with 36 healthy volunteers, CCK-4 (0, 9, 25 and 50 μg) was also found to induce panic attacks in a dose-dependent manner; the panic rate was 11% (9 μg), 17% (25 μg) and 47% (50 μg).[15] No panic attacks occurred with placebo injections.

### Antipanic agents should block the effects of the agent

Recently, we have demonstrated that the panicogenic effects of CCK-4 can be antagonized by chronic treatment with imipramine.[16] Specifically, 11 patients with panic disorder who displayed a positive panicogenic response to CCK-4 (20 μg) were treated with imipramine on a chronic basis and rechallenged with CCK-4 (20 μg) after being free of both panic and agoraphobic symptoms for at least 8 weeks. With rechallenge, patients displayed a marked reduction in the number and sum intensity of symptoms, duration of symptoms and cardiovascular responsiveness. Moreover, only 2 of the 11 patients who previously panicked with CCK-4 experienced a panic attack when rechallenged. It is also notable that patients who consumed higher doses of imipramine experienced fewer and less intense panic symptoms at rechallenge, suggesting that the decreased sensitivity to CCK-4 following chronic imipramine therapy was most likely attributed to a drug effect rather than to other factors such as spontaneous remission of symptoms.

In addition to our study with imipramine, there is evidence that treatment with another antipanic agent can inhibit the effects of CCK-4 in humans. Van Megen and colleagues have reported that the selective serotonin reuptake inhibitor, fluvoxamine, decreased CCK-4-induced panic attacks in patients with panic disorder.[17]

### The effects of the agent are not antagonized by drugs without antipanic effects

There is indirect evidence that CCK-4 also satisfies this criterion. In the context of investigating the effects of CCK-B receptor antagonists on CCK-4-induced panic symptoms, we observed that pretreatment with placebo failed to antagonize CCK-4-induced panic symptoms in patients with panic disorder.[14] In another study which investigated the possible mediating role of benzodiazepine receptors in CCK-4-induced panic symptoms, pretreatment with the benzodiazepine receptor antagonist flumazenil, a compound without any known antipanic activity, failed to diminish response to CCK-4 challenge in healthy volunteers.[18]

## COMPARISON OF THE EFFECTS OF CCK-4 TO THOSE PRODUCED BY $CO_2$

Another research approach we have employed in evaluating whether CCK-4 is a valid panicogen has been to compare its effects with those produced by other valid pharmacological models of panic. So far, we have compared response to a 25 µg dose of CCK-4 and a single inhalation of 35% $CO_2$ in patients with panic disorder[19] and healthy volunteers.[20] In the study with patients, CCK-4 produced a greater number of symptoms and more intense symptoms than 35% $CO_2$. CCK-4 was also more effective in inducing panic attacks than $CO_2$ (91% versus 45%), although the profile of symptoms which emerged in response to either agent was similar in patients who experienced a panic attack. While CCK-4 was found to produce more intense panic symptoms than $CO_2$ in healthy volunteers, these concentrations of CCK-4 and $CO_2$ were equipotent in promoting panic attacks (17% versus 21%). It will be interesting in future studies to compare CCK-4 with other panicogenic challenges, particularly the frequently employed sodium lactate infusion.

Overall, the data just summarized demonstrate that CCK-4 satisfies previously established criteria for an ideal panicogenic agent and that it compares well to at least one widely accepted pharmacologic model of panic. It is also important to mention that of all pharmacological agents known to provoke panic attacks in humans, including sodium lactate, $CO_2$, caffeine, yohimbine, isoproterenol and mCPP, CCK-4 is the only one that fulfills criteria for a neurotransmitter. CCK is well characterized in the CNS and is abundant in brain regions implicated in the promotion of panic attacks including the brainstem, hippocampus, amygdala and cerebral cortex.[21] Moreover, biochemical and electrophysiological data suggest interactions between CCK and multiple neurotransmitter systems, including serotonin, noradrenaline, GABA and dopamine. As a panicogenic agent, therefore, CCK-4 provides an important opportunity to identify an endogenous anomaly associated with panic disorder, and to enhance our understanding of the multiple neurotransmitter systems which potentially contribute to the generation of panic attacks.

Another important feature of CCK-4 is that it is simple to administer in a low volume intravenous bolus infusion (in less than 5 seconds). This method of administration has considerable advantage over the slow infusion procedures required to induce symptoms of panic with other panicogens, particularly sodium lactate. The relatively protracted infusion interval has been associated with physiologic alterations, such as volume overload, and metabolic changes that can introduce non-specific psychological effects.[22] Another technical advantage is that the latency to effect symptoms of panic with CCK-4 is rapid and predictable, permitting measurement of central and peripheral nervous system activity during the interval associated with peak panic symptoms. Considered together, the technical advantage of CCK-4 administration, coupled by its presence in the CNS, commends its use for research into the pathophysiology of panic disorder.

The data generated from our validation studies also highlight some of the usefulness of CCK-4 as a panicogenic challenge for research in anxiety. For instance, demonstrating that the effects of CCK-4 are reproducible in the same patient has important implications in relation to testing the effectiveness of antipanic drugs in blocking CCK-4-induced panic symptoms. In addition, the dose-response study in patients indicated that a 20 μg dose of CCK-4, which produced panic attacks in 75% of patients, might be suitable for efficacy studies. In particular, this dose promotes noticeable changes in behavior and other indices of anxiety without being excessively potent to mask the effectiveness of potential antipanic drugs to block the effects of CCK-4. We have successfully used this dose to evaluate the effects of CCK-B receptor antagonists and imipramine on CCK-4-induced panic symptoms.

## MECHANISM OF ANXIOGENIC ACTION OF CCK-4

The mechanism and localization of CCK-4-induced panic symptoms are still largely unknown. Several investigations using animal models of anxiety have revealed that the anxiogenic effects of CCK-4 are blocked by selective CCK-B receptor antagonist, suggesting that CCK-B receptors are an important site of anxiogenic action of exogenous CCK-4. We have completed two studies which suggest that CCK-B receptors are also important mediators of the behavioral and cardiovascular changes during CCK-4 challenge in humans.

In one study, patients with panic disorder were pretreated with L-365,260 (10 or 50 mg po) or placebo 90 minutes before challenge with CCK-4.[23] Analysis of the data indicated that the 50 mg dose of L-365,260 was superior over placebo in reducing the number of symptoms induced with CCK-4, the sum intensity of symptoms and panic attack frequency. Moreover, when compared with placebo, both the 10 and 50 mg dose of the antagonist dramatically decreased CCK-4-evoked increases in heart rate. In the other study, we have evaluated the effects of another CCK-B antagonist, CI-988 (100 mg), on CCK-4-induced panic symptoms in 30 healthy males in a placebo-controlled, double-blind design.[24] There was a significant decrease in sum intensity scores and panic attack frequency following CI-988 treatment.

These studies suggest that CCK-B receptors are mediators of the panicogenic effects of CCK-4 in humans. In addition to investigating the involvement of CCK-B receptors, we have evaluated whether benzodiazepine receptors contribute to the behavioral effects of CCK-4. As indicated earlier, benzodiazepine receptor agonists selectively and specifically antagonized CCK-8S-induced excitation of rat hippocampal neurons. It was subsequently demonstrated that neuronal responsiveness to CCK-8 in rats decreases following long-term administration of benzodiazepine agonists.[25] There is also evidence that benzodiazepine receptor agonists attenuate the anxiogenic effects of exogenous CCK-4 in nonhuman primates[26] and healthy volunteers,[5] although such data do not establish that benzodiazepine receptors invariably contribute to the panicogenic effects of CCK-4.

To further explore the role of benzodiazepine receptors in CCK-4 induced panic symptoms we determined whether pretreatment with the benzodiazepine receptor antagonist flumazenil could influence response to CCK-4 in healthy volunteers, using a double-blind placebo control cross-over design. In investigating the potential interaction between CCK and benzodiazepine receptors in CCK-4-induced panic response, we proposed a model that was based on the premise that exogenous CCK-4, through actions on

CCK-B receptors, interact with benzodiazepine receptors in eliciting symptoms by indirectly acting like a benzodiazepine receptor inverse agonist. In other words, we postulated that CCK-4 might act as an endogenous "virtual" inverse agonist of benzodiazepine receptors. To support this hypothesis, it was necessary to determine whether flumazenil could antagonize the panicogenic effects of CCK-4. Our findings indicate that benzodiazepine receptors are not mediators of CCK-4-induced panic symptoms in normal subjects. We found no discernible difference between flumazenil and placebo pretreatment for the number of symptoms induced with CCK-4, sum intensity of symptoms and panic attack frequency.[18]

The mechanism by which the NE, 5-HT and adenosine systems participate in CCK-4 induced anxiety has yet to be investigated. Our study which showed that chronic treatment with imipramine, which inhibits NE and 5-HT reuptake, could antagonize the panicogenic effects of CCK-4 in patients with panic disorder argue that these monoamines may be instrumental in interacting with CCK in promoting symptoms of panic. There are also some interesting data suggesting that 5-$HT_3$ receptors are mediators of CCK-induced anxiety. In this respect, Vasar and his associates[27] reported that the anxiogenic effect of caerulein, a CCK agonist, was prevented by prior treatment with the 5-$HT_3$ receptor antagonist ondansetron. A possible interaction between CCK and adenosine receptors is supported by the finding that NECA (N-ethycarboxamido-adenosine), an agonist of adenosine $A_2$ receptors, antagonized the anxiogenic effects of CCK-4 in nonhuman primates.[26]

Another question of central importance concerns the site(s) of action of this peptide in humans. Currently, there is no available evidence that CCK-4 crosses the blood-brain barrier, although the possibility exists that CCK-4 affects CCK-B receptors in brain regions which are not fully protected by the blood-brain barrier. Knowledge of cardiovascular neurophysiology, as well as studies of the behavioral and cardiovascular effects of CCK-4, permit some speculation as to the possible site of action of CCK-4. Some investigators have suggested that brainstem regions ordinarily function to monitor sympathetic nerve discharge and vasomotor tone. Increases in blood pressure and heart rate have been observed following electrical or pharmacological stimulation of the nucleus tractus solitarius (NTS),[28] the medullary nuclei[29,30] and the parabrachial nucleus.[31] These brainstem regions are inter-related by diverse neuronal projections and are connected to adrenergic structures,[32,33] such as the locus ceruleus, which are postulated to play a role in panic attacks.[34] Further, there is experimental data suggesting that CCK interacts with these brainstem mechanisms in modulating respiratory and cardiovascular functions. Microiontophoretic application of CCK-8S to neurons of the NTS decreased both neuronal firing and respiratory frequency in cats; effects which were reversed by administration of CCK-4.[35]

Our clinical investigations have demonstrated that exogenous CCK-4 produces robust and dose-dependent increases in heart rate and blood pressure.[14] Moreover, pretreatment with the CCK-B antagonist L-365,260 significantly decreased CCK-4-induced increases in heart rate.[23] It might be argued that increases in cardiovascular activity in response to CCK-4 challenge may be the result of direct or indirect stimulation of CCK receptors in brainstem structures such as the NTS. It is also conceivable that the evocation of emotional and psychosensorial symptoms following CCK-4 challenge results from an action of CCK-4 on brainstem structures and a subsequent activation or inhibition of higher CNS regions mediated by neuronal projections. As these brainstem structures are not fully shielded by the blood-brain barrier, CNS penetration by CCK-4 might not even be necessary for this action. This might also explain the rapid (in less than 1 minute) appearance of symptoms observed in both panic disorder and normal control subjects following CCK-4 challenge.

## CONCLUSION

This chapter has summarized the data supporting the validity of CCK-4 as a panicogenic agent and the hypothesis that alterations in CCK activity might be involved in the pathogenesis of panic disorder. It remains to be determined whether the experience of panic anxiety is related to an endogenous malfunction of the CCK system. Conventional approaches, such as measurement of body fluid concentrations of metabolites, might be helpful in answering this question. It is notable in this respect that Lydiard and his colleagues[36] reported that CSF concentrations of CCK-8S were markedly decreased in panic patients relative to control subjects. Also, Brambilla and colleagues have reported decreased concentrations of CCK-8S in lymphocytes of panic patients as compared to controls.[37] These results suggest that panic disorder might be due to an abnormal production or turnover of CCK-8S or CCK-4. This hypothesis can be systematically tested once more sensitive analytical techniques which permit measurement of shorter CCK fragments, like CCK-4, are available. Nevertheless, the finding that panic patients, relative to control subjects, have decreased concentrations of CCK-8S, a mixed CCK-A/CCK-B receptor agonist, and exhibit a more severe panic response to CCK-4, a selective CCK-B receptor agonist, suggests that panic attacks may result from an imbalance between CCK-A and CCK-B receptor systems. Admittedly speculative, such a hypothesis is deserving of further consideration particularly in view of recent findings that CCK-A and CCK-B agonists have opposite effects on NTS neurons[38] and on rodent behavior.[39]

An important step in testing the hypothesis that panic disorder is the result of an endogenous anomaly of the CCK-B system will be pharmacological validation. If spontaneous panic attacks are the result of an enhanced activity of the CCK-B system, then CCK-B antagonists should have antipanic effects. This question remains unanswered as the two CCK-B antagonists, L-365,260 and CI-988, which have been used in clinical trials so far are limited by their unfavorable pharmacokinetic properties.[24,40] Fortunately, clinical research with CCK-4 has received considerable attention from industry. Several pharmaceutical companies have recently developed other CCK-B receptor antagonists and have put these agents on fast-track drug development programs with the intention of testing their efficacy in panic and other anxiety disorders. It should be emphasized that even if studies on the pharmacological validation of the CCK hypothesis of panic yield negative results, the CCK-4 challenge paradigm could nevertheless remain a practical research tool that can enhance our knowledge of the neurobiological mechanisms subserving panic attacks and lead to the development of novel antipanic drugs.

## ACKNOWLEDGMENTS

The clinical studies reviewed in this paper received financial support from the Fonds de la Recherche en Santé du Québec, the Medical Research Council of Canada, St. Mary's Hospital Foundation and Psychopharmacology Fund. Thanks are extended to Lynnette Ashton for secretarial assistance.

## REFERENCES

1. Bradwejn J, de Montigny C. Benzodiazepines antagonize cholecystokinin-induced activation of rat hippocampal neurons. Nature 1984; 312:363-364.
2. Rehfeld J F. CCK and anxiety: Introduction. In: eds Dourish CT, Cooper SJ, Iversen SD, Iversen LL. Multiple Cholecystokinin Receptors in Man. 1992:117-120. Oxford University Press, New York.
3. Vanderhaeghen J, Crawley J. Neuronal Cholecystokinin. Ann NY Acad Sci 1985; 448:1-697.
4. Bradwejn J, Koszycki D, Meterissian G. Cholecystokinin-tetrapeptide induced panic attacks inpatients with panic disorder. Can J Psychiatry 1990; 35:83-85.
5. de Montigny C. Cholecystokinin tetrapeptide induces panic-like attacks in healthy volunteers: Preliminary findings. Arch Gen Psychiatry 1989; 46:511-517.
6. Guttmacher LB, Murphy DL, Insel TR. Pharmacologic models of anxiety. Comprehensive Psychiatry 1993; 24:312-326.
7. Gorman J M, Fyer MR, Liebowitz MR. Pharmacologic provocation of panic attacks. In Psychopharmacology: A Third Generation of Progress. Meltzer HY, ed. 1987:980-983. Raven Press. New York, NY.
8. Bradwejn J, Koszycki D. CCK-4 and panic attacks in man. In eds Dourish CT, Cooper SJ, Iversen SD, Iversen LL. Multiple Cholecystokinin Receptors in Man. 1992:121-131. Oxford University Press, Oxford, England.
9. Bradwejn J, Koszycki D, Shriqui C. Enhanced sensitivity to cholecystokinin-tetrapeptide in PD: Clinical and behavioral findings. Arch Gen Psychiatry 1991; 48:603-607.
10. Abelson JL, Nesse RM. Cholecystokinin-4 and panic. Arch Gen Psychiatry 1990; 47:395.
11. Abelson JL, Nesse RM. Pentagastrin infusions in patients with panic disorder. I Symptoms and cardiovascular responses. Biol Psychiatry 1994; 36:73-83.
12. van Megen HJM, den Boer JA, Westenberg HGM. Single blind dose response study with cholecystokinin in panic disorder. Clin Neuropharm 1992; 15(suppl 1):532B.
13. Bradwejn J, Koszycki D, Payeur R, Bourin M. Study of the replication of action of cholecystokinin in panic disorders. Am J Psychiatry 1992; 149:962-964.
14. Bradwejn J, Koszycki D, Annable L. A dose-ranging study of the behavioral and cardiovascular effects of CCK-tetrapeptide in panic disorder. Biol Psychiatry 1992; 32:903-912.
15. Bradwejn J, Koszycki D, Bourin M. Dose ranging study of the effect of $CCK_4$ in healthy volunteers. J Psychiatry Neurosci 1991; 16:260-264.
16. Bradwejn J, Koszycki D. Imipramine antagonizes the panicogenic effects of CCK-4 in panic disorder patients. Am J Psychiatry 1994; 151:261-263.

17. van Megen H, Westenberg H, den Boer J. Effect of the selective serotonin reuptake inhibitor (SSRI) fluvoxamine on CCK-4 induced panic attacks (Abstract) Neuropsychopharmacol 1994; 10:270S.
18. Bradwejn J, Koszycki D, Couetoux du Tertre A, Paradis M, Bourin M. Effects of flumazenil on cholecystokinin-tetrapeptide-induced panic symptoms in healthy volunteers. Psychopharmacol 1994; 114:257-261.
19. Bradwejn J, Koszycki D. Comparison of $CO_2$-induced panic attacks with cholecystokinin-induced panic attacks in PD. Prog Neuro-Psychopharmacol Biol Psychiat 1991; 15:237-239.
20. Koszycki D, Bradwejn J, Bourin M. Comparison of the effects of cholecystokinin and carbon dioxide in healthy volunteers. Eur Neuropharmacol 1991; 1:137-141.
21. Karkanias CD, Block GA, Reines S et al. Neurobiology of panic disorder. Letter, Am J Psychiatry 1989; 146:1357. .
22. Margraf J, Ehlers A, Roth WT. Sodium lactate infusions and panic attacks: A review and critique. Psychosom Med 1986; 48: 23-51.
23. Bradwejn J, Koszycki D, Couetoux-Dutertre A, van Megen H, den Boer J, Westenberg H, Annable L. L-365,260: a CCK-B antagonist blocks CCK-4-panic in panic disorder. Arch Gen Psychiatry 1994; 51:486-493.
24. Bradwejn J, Koszycki D, Paradis M, Reece P, Sedman R. The effects of CI-988 on CCK-4-induced panic in healthy volunteers. Biol Psychiat In Press.
25. Bouthillier A, de Montigny C. Long term benzodiazepine treatment reduces neuronal responsiveness to cholecystokinin: an electrophysiological study in the rat. Eur J Pharmacol 1988; 115:135-138.
26. Palmour R, Bradwejn J, Ervin F. The anxiogenic effects of $CCK_4$ in monkeys are reduced by CCK-B antagonists, benzodiazepines or adenosine $A_2$ agonists. Clin Neuropharmacol 1992; 15(Suppl. 1):489B.
27. Vasar E, Peuranen E, Ööpik T et al. Ondansetron, and antagonist of $5HT_3$ receptors, antagonizes the anti-exploratory effect of caerulein, an agonist of CCK receptors, in the elevated plus maze. Psychopharmacol 1993; 110:213-218.
28. Jordan D, Spyer KM. Brainstem integration of cardiovascular and pulmonary afferent activity. Prog Brain Res 1986; 67:295-314.
29. Dampney RAL, Goodchild AK, Robertson LG et al. Role of ventrolateral medulla in vasomotor regulation: A correlative anatomical and physiological study. Brain Res 1982; 249: 223-235.
30. Pilowsky R, West M, Chalmers J. Renal sympathetic nerve responses to stimulation, inhibition and destruction of the ventrolateral medulla in the rabbit. Neurosci Lett 1985; 60:51-55.
31. Marovitch S, Kumada M, Reis DJ. Role of parabrachialis in cardiovascular regulation in the cat. Brain Res 1982; 232:57-75.

32. Dampney RAL, Czachurski J, Dembowsky K et al. Afferent connections and spinal projections of the pressor region in the rostral ventrolateral medulla of the cat. J Auton Nerv Syst 1977; 20:73-86.
33. Ross CA, Ruggiero DA, Park DH et al. Tonic vasomotor control by the rostral ventrolateral medulla: Effect of electrical or chemical stimulation of the area containing C1 adrenaline neurons on arterial pressure, heart rate, and plasma catecholamines and vasopressin. J Neurosci 1984; 4:474-494.
34. Gorman JM, Liebowitz MR, Fyer AJ et al. Neuro-anatomical hypothesis for panic disorder. Am J Psychiatry 1989; 146:148-161.
35. Denavit-Saubié M, Hurlé MA, Morin-Surun MP et al. The effects of cholecystokinin-8 in the nucleus tractus solitarius. In: eds. Vanderhaeghen JJ, Crawley JN. Neuronal Cholecystokinin. Ann NY Acad Sci 1985:375-384, New York, New York.
36. Lydiard B, Ballenger J, Laraia M et al. CSF cholecystokinin concentrations in patients with panic disorder and normal comparisons subjects. Am J Psychiatry 1992; 149:691-693.
37. Brambilla F, Bellodi L, Perna G, Garberi A, Panerai A, Sacerdote P. Lymphocyte cholecystokinin concentrations in panic disorder. Am J Psychiatry 1993; 150:1111-1113.
38. Branchereau P, Böhme GA, Champagnat J et al. Cholecystokinin$_A$ and cholecystokinin$_B$ receptors in neurons of the brainstem solitary complex of the rat: pharmacological identification. J Pharmacol Exp Ther 1992; 260:1433-1440.
39. Männistö PT, Lang A, Harro J, Peuranen E, Bradwejn J, Vasar E. Opposite effects mediated by CCKA and CCKB receptors in behavioural and hormonal studies in rats. Naunyn-Smiedeberg's Arch Pharmacol 1994; 349:478-484.
40. Kramer MS, Cutler NR, Ballenger JC, Patterson WM, Mendels J, Chenault A, Shrivastava R, Matzurah-Wolfe D, Lines C, Reines S. A placebo-controlled trial of L-365,260, in panic disorder. Biol Psychiatry 1995; 37:462-466.

CHAPTER 5

# Psychological Factors and Response to Cholecystokinin

**Diana Koszycki**

## INTRODUCTION

Various pharmacological agents (e.g., carbon dioxide ($CO_2$), sodium lactate, yohimbine) are known to provoke symptoms associated with panic disorder in human subjects. Moreover, there are marked interindividual differences in behavioral sensitivity to these agents. In particular, the effectiveness of these agents in provoking panic attacks is more evident in panic disorder patients relative to healthy volunteers (for review see ref. 1). Although variation in behavioral sensitivity to panicogens is believed to reflect alterations in the neurochemistry instrumental to the expression of clinical panic attacks, some investigators have argued that psychological variables including anticipatory anxiety, interoceptive sensitivity, appraisal of threat or harm, perception of control and panic expectancy, among other variables, play a more salient role in the induction of anxious/fearful feelings (for review see refs. 2 and 3).

Systemic administration of the central cholecystokinin (CCK-B) receptor agonist, CCK-4, can produce symptoms reminiscent of clinical panic attacks in both patients with panic disorder and individuals without a personal or family history of panic.[4,5] The frequency of CCK-4 induced panic is more conspicuous among panic patients relative to healthy subjects,[5] suggesting that the disorder may be associated with alterations in CCK neurotransmission. Such an interpretation has received some support from the demonstration that panic patients exhibit decreased CSF[6] and lymphocyte[7] CCK-8 concentrations compared to normal subjects. However, it is also conceivable that psychological factors are a crucial determinant of behavioral sensitivity to CCK-4 challenge. In the present chapter, the author will examine the available data describing the contribution of psychological factors to the induction of panic anxiety states following systemic CCK-4 administration. It should be mentioned at the outset that consideration of the role of psychological processes in CCK-induced panic anxiety has only recently been introduced and firm conclusions are not currently available.

## COGNITIVE ATTRIBUTION

Cognitive models of panic posit that the subjective experience of fear and anxiety during a panic attack occurs because somatic sensations resulting from autonomic arousal (e.g., palpitations, choking, shaking/trembling) are systematically appraised as having catastrophic consequences such as death, illness or loss of control of emotions and/or behavior.[8] This model assumes that cognitive factors are etiologically significant for panic disorder, rather than just a symptom of the illness—a supposition which has yet to be empirically verified. The tenability of a purely cognitive explanation of the experience of fear and anxiety during a panic attack has been challenged by the demonstration that panic attacks occur during non-REM sleep,[9,10] during relaxation training[11] and in the absence of fearful cognitions.[12,13] Whatever the explanation for these latter phenomena, it is reasonable to suppose that catastrophic appraisal of somatic sensations is sufficient but not essential in the provocation of panic anxiety states in clinical settings.

In addition to according a causal role to cognitions in the onset and maintenance of clinical panic attacks, cognitive theorists have postulated that catastrophic appraisal of somatic symptoms induced by panicogenic agents is salient to the provocation of fear and anxiety.[2,3] A core concept in this framework is that panicogenic substances, while capable of eliciting the somatic symptoms of panic, are not in and of themselves sufficient to provoke fear and anxiety. Only when they interact with self-referent thoughts that reflect themes of harm and danger is the affective manifestation of panic thought to occur.[8] Cognitive theorists assume that the enhanced sensitivity to panicogenic agents evident in panic patients is attributable to an enduring tendency for these patients to interpret bodily sensations in a catastrophic fashion. However, empirical data supporting this view are at present limited and overall results are mixed and inconclusive. For instance, whereas some investigators reported that the provision of explicit information for bodily sensations elicited by $CO_2$[14] and sodium lactate[15] challenges mitigated the experience of fear and anxiety in patients with panic disorder, other investigators using this same paradigm failed to detect demonstrable decreases in anxiety ratings and panic frequency in patients undergoing a series of respiratory challenges (e.g., 5% and 7% $CO_2$).[16] Further, it has been noted that panic patients display greater reactivity than controls to sleep lactate infusions[17] and that the pharmacological induction of aversive and threatening somatic symptoms does not invariably provoke panic in patients.[18] Collectively, these findings suggest that the association between catastrophic misinterpretation and the induction of panic anxiety states following administration of panicogenic substances is equivocal at best.

Is there a link between catastrophic thoughts and the emergence of anxious/fearful affect following systemic administration of CCK-4? Two lines of evidence suggest that this link is weak. First, all subjects who participate in our CCK challenge studies are provided with explicit information about the possible effects of CCK-4. In addition, they are instructed to attribute the somatic symptoms they experience to the transient effects of the peptide. According to cognitive theory, these instructions should markedly attenuate the degree of anxiety experienced (i.e., subjects are provided with a rational and noncatastrophic explanation for their symptoms). However, such a theory does not appear to be supported. The provision of detailed information about the pharmacological effects of CCK-4 has been found to neither detract from the expression of a panic attack nor limit the degree of anxiety experienced in panic patients or even in healthy volunteers. For instance, Bradwejn et al[5] reported that systemic administration of a 50 μg dose of CCK-4 provoked panic episodes in 100% of panic patients and in 47% of healthy volunteers. Moreover, CCK-4 engendered severe or extremely severe fear and anxiety in all of the patients and in 86% of the control subjects in whom a panic attack was transiently induced.

Second, we have noted that a subgroup of panic disorder and normal control subjects who are challenged with CCK-4 experience "noncognitive" panic attacks.[19,20] That is, they experience both the somatic and affective components of panic anxiety, but none of the associated catastrophic thoughts (e.g., fear of dying, fear of losing control, fear of going crazy). We recently compared the effects of CCK-4 challenge in subjects who panicked but experienced no catastrophic cognition (non-cognitive panickers) and in those who panicked and experienced at least one catastrophic cognition (cognitive panickers).[21] In normal subjects challenged with a 50 μg dose of CCK-4, no significant difference in affective intensity emerged as a function of panic subtype. Further, the correlation between the intensity of affect and the number of cognitive symptoms endorsed was small and nonsignificant. Analysis of somatic symptoms of panic revealed that the overall pattern of symptom endorsement was comparable for cognitive and noncognitive panickers. In patients with panic disorder challenged with a 25 μg dose of CCK-4, neither the severity of affective response nor the severity of somatic response to CCK-4 was found to distinguish between cognitive and noncognitive panickers.

Summarizing, the aforementioned data suggest that the cognitive process of self-labeling is not a major causal influence on affective response to systemic CCK-4. The observation that the expression of fear and anxiety following CCK-4 administration is, in some cases, independent of cognitive processing suggests that the enhanced sensitivity to the pharmacological effects of CCK-4 evident in panic patients cannot be simply ascribed to a tendency for these patients to associate somatic sensations with dramatic outcomes. Although CCK-4 does engender cognitive changes in the majority of panic patients tested, it is the author's impression that such cognitive alterations are a consequence rather than a cause of CCK-4 induced panic anxiety. Nevertheless, further empirical research is needed to sort out the temporal and causal sequence of affect and cognitions. Finally, it would be of considerable interest to investigate the neurochemical and neuroanatomical concomitants of cognitive and noncognitive panic attacks provoked by CCK-4. Such an investigation could provide important insights into the neurobiological mechanisms underlying such cognitive functions as attention, memory and anticipatory cognitive activity in panic disorder.

## ANXIETY SENSITIVITY

Anxiety sensitivity is a prominent characteristic of patients with panic disorder. It is defined as a stable dimension of personality that consists of fears of physical sensations based on the belief that these symptoms have harmful consequences.[22] Anxiety sensitivity is not dependent on a prior history of panic attacks, but rather, it is hypothesized to be an important cognitive risk factor in the development of panic attacks and panic disorder. This hypothesis is supported by some[23,24] but not all[25] studies using nonclinical subjects.

It has been proposed that anxiety sensitivity is an important psychological variable that mediates response to panicogenic stimuli. That is, individuals who fear the somatic symptoms produced by such stimuli respond to them with greater fear and anxiety. Given that panic patients exhibit high levels of anxiety sensitivity relative to normal controls, it might be expected that they would react more anxiously to the induction of arousal symptoms. Studies that attempt to show a relationship between anxiety sensitivity and behavioral reactivity to panic provocation procedures are few and inconclusive. Holloway and McNally[26] tested response to voluntary hyperventilation, a relatively minor aversive event, in college students with high and low self-ratings of anxiety sensitivity and found that individuals with high anxiety sensitivity responded more anxiously to the challenge than those with low anxiety sensitivity. However, this finding may have been due to differences in baseline anxiety. Indeed, adjusting for differences in baseline anxiety scores resulted in a sizable drop in the p value to non-significant levels. It should be noted as well that the hyperventilation challenge effected a relatively small increase in self-rated anxiety and none of the subjects panicked while hyperventilating. This suggests that anxiety sensitivity may well be associated with anticipatory anxiety rather than anxiety associated with panic attacks. Rapee and colleagues[27,28] reported that anxiety sensitivity emerged as a significant predictor of affective response to hyperventilation and 5.5% $CO_2$ challenges, although the amount of variance accounted for by anxiety sensitivity was not impressive. For instance, anxiety sensitivity accounted for only 8% of the variance in affective response to $CO_2$ inhalation in a combined sample of panic disorder and control subjects.[26] This suggests that factors other than anxiety sensitivity are likely to be involved in the mediation of affective response to panicogenic stimuli.

Our laboratory has recently addressed the question of whether individual differences in anxiety sensitivity, as measured by the Anxiety Sensitivity Index,[22] play a crucial role in mediating behavioral sensitivity to CCK-4 challenge. In a preliminary study of 36 healthy volunteers,[19] systemic administration of a 50 μg dose of CCK-4 provoked a comparable profile of anxiety and panicogenic response in individuals with low, medium (i.e., average) and high levels of anxiety sensitivity, as well as similar increases in heart rate and blood pressure. The frequency of CCK-4-induced panic attacks was 56% for subjects with low anxiety sensitivity, 65% for those with medium anxiety sensitivity and 60% for those with high anxiety sensitivity. Interestingly, although subjects with high anxiety sensitivity were not more predisposed to experiencing a panic attack with CCK-4 than the other subjects, they were more likely to report catastrophic thoughts. This suggests that anxiety sensitivity may be related to interpretation of symptoms rather than to mediation of panic. Nevertheless, it might be noted that in our analysis of cognitive and noncognitive panickers, discernible differences in self-ratings of anxiety sensitivity were not detected.[21] Accordingly, it is unlikely that the emergence of panic cognitions is solely determined by elevated anxiety sensitivity.

Although our preliminary data suggest that the frequency of CCK-4-induced panic attacks does not vary with level of anxiety sensitivity, there is still some question as to whether the quality of panic induced by CCK-4 is modified by level anxiety sensitivity. To address this question, we evaluated the symptom profile associated with CCK-4-induced panic in subjects with low, medium and high levels of anxiety sensitivity.[29] Preliminary analysis of data obtained from 64 healthy volunteers in whom a panic attack was transiently induced revealed that the intensity of affective response to CCK-4 was identical for the three anxiety sensitivity groups. In addition, inspection of DSM-IV somatic and cognitive symptoms of panic revealed that the symptom endorsement rates were comparable across the three groups. However, intensity ratings for dizziness, chest/pain discomfort, and fear of dying were higher in high anxiety sensitivity subjects relative to the other subjects. Interestingly, whereas high anxiety sensitivity was associated with higher symptom rating for fear of dying, low anxiety sensitivity was associated with a higher (albeit nonsignificant) symptom intensity rating for fear of losing control.

We have also examined the extent to which anxiety sensitivity influences the magnitude of response to CCK-4 in patients with panic disorder.[20] In this study, anxiety sensitivity did not correlate with the intensity of affective response to CCK-4, although it did correlate with the number and severity of cognitive symptoms at a modest level ($r=.32$ and $r=.36$, respectively, $p$'s$<0.05$). Inspection of individual cognitive symptoms revealed that anxiety sensitivity correlated significantly with fear of losing control ($r=.46$, $p<0.01$). On the other hand, the relationships with fear of going crazy and fear of dying were not significant ($r=.24$ and $r=.13$, respectively). There was no association between anxiety sensitivity and the number of somatic symptoms endorsed or the sum intensity of somatic symptoms.

In summary, the results from these studies do not provide compelling evidence that patients with panic disorder exhibit an enhanced sensitivity to the panicogenic effects of CCK-4 relative to normal controls simply because they possess high levels of anxiety sensitivity. Nevertheless, it remains to be determined whether response to decreasing CCK-4 doses is influenced by self-rated anxiety sensitivity in normal subjects. Indeed, it is conceivable that differences between low, medium and high anxiety sensitivity groups may be observed only with lower dosages of CCK-4. Finally, the apparent discrepancy between the results with CCK-4 and those with hyperventilation in healthy volunteers suggests that the association between anxiety sensitivity and behavioral reactivity to challenges may depend, in part, on the nature and potency of the panicogenic stimulus employed. Voluntary hyperventilation has unreliable anxiogenic effects and is considered by some authors to be a poor model of panic.[30] In contrast, CCK-4 produces a reliable, dose dependent profile of symptoms reminiscent of panic attacks and fulfills criteria for an ideal panicogenic agent.[31]

## PERSONALITY CORRELATES

As noted, the magnitude of behavioral response to systemic CCK-4 administration in patients with panic disorder does not appear to be strongly influenced by individual differences in anxiety sensitivity. However, it is conceivable that other personality traits play a role. Accordingly, we determined whether personality traits that presumably underlie anxiety disorders amplify behavioral sensitivity to systemic CCK-4 administration in patients with panic disorder.[20] The standard MMPI[32] (Minnesota Multiphasic Personality Inventory) clinical scales measuring Hypochondriasis, Depression, Hysteria, Psychasthenia and Social Introversion and the supplemental MMPI Anxiety scale[33] were used to measure these dispositions.

Correlational analyses failed to detect significant relationships between somatic, cognitive and affective responses to CCK-4 and scores on the MMPI Hypochondriasis, Depression, Hysteria and Psychasthenia scales. By contrast, significant correlations in the moderate range were found between the MMPI Social Introversion scale and the number and sum intensity of somatic symptoms ($r=.37$ and $r=.41$, $p$'s$<0.05$), the number and sum intensity of cognitive symptoms ($r=.42$, $p<0.05$ and $r=.50$, $p<0.01$) and the intensity of anxiety/fear/apprehension ($r=.58$, $p<0.001$). The somatic symptoms most strongly associated with the Social Introversion scale were trembling/shaking ($r=.54$, $p<0.001$) and feeling detached/unreal ($r=.44$, $p<0.01$). The cognitive symptoms most strongly associated with the social introversion scale were fear of going crazy ($r=.44$, $p<0.001$), fear of dying ($r=.40$, $p<0.05$) and fear of losing control ($r=.39$, $p<0.05$). The MMPI Anxiety scale also showed a modest correlation with the number and sum intensity of cognitive symptoms ($r=.35$ and $r=.37$, respectively, $p$'s$<0.05$). The cognitive symptom most strongly associated with this scale was fear of dying ($r=.36$, $p<0.05$).

To further clarify the relationship between personality traits and behavioral response to CCK-4, hierarchical multiple regression analyses were conducted. The predictor variables that were included in the regression analyses included the Anxiety Sensitivity Index, the MMPI Hypochondriasis scale and the MMPI Social Introversion scale. The Anxiety Sensitivity Index, which was entered into the model first, was not found to be a significant predictor of the magnitude of response to CCK-4, as measured by the three anxiety subscales. Further, the addition of the MMPI Hypochondriasis scale did not add significantly to the model for any of the measures. In contrast, the MMPI Social Introversion scale, which was entered next, contributed significantly and meaningfully to the predictive value. The magnitude of change was greatest for the intensity of affective response to CCK-4, followed by the intensity of catastrophic cognitions the number of reported catastrophic cognitions, and the intensity of somatic symptoms.

In summary, the aforementioned data suggest that the behavioral consequences of CCK-4 in patients with panic disorder are largely independent of personality variables. The only notable exception is that the MMPI Social Introversion scale, a measure of neurotic introversion, was a significant predictor of behavioral response to CCK-4. The variables which potentially account for the apparent enhanced sensitivity of socially introverted patients to the behavioral consequences of CCK-4 are not readily available, although some provisional hypotheses may be proposed. It is conceivable that endogenous variations of central and/or peripheral CCK activity may be a neurochemical concomitant of the introversion personality trait. This argument deserves some consideration in view of the recent identification of an negative correlation between plasma gastrin, which binds preferentially to CCK-B receptors, and measures of socialization in patients with functional gastrointestinal disorders[34] and in healthy women.[35] It is also interesting to note in this regard that the social isolation of rats leads to upregulation of CCK receptors in the frontal cortex.[36] Accordingly, it is conceivable that the enhanced sensitivity to CCK-4 observed in socially introverted panic patients is attributed, at least in part, to variation in CCK receptor function. It may also be considered that a highly tenuous link between CCK and introversion may be found in the demonstration of CCK availability and a CCK receptor distribution in the septophippocampal and ascending reticular activating system. The provisional involvement of these central structures and CCK in introversion are predicated upon the speculative neuroanatomical hypothesis provided by Gray[37] and Eysenck,[38] respectively. Clearly any conclusions drawn from such an analysis would require elaborate clinical verification.

## BASELINE ANXIETY

A major issue in the study of pharmacologically-induced panic anxiety is the potential confounding effects of baseline anxiety. Margraf and colleagues[39] have argued that baseline anxiety and autonomic arousal prime patients to panic during biological challenge studies. Further, they posit that enhanced sensitivity to panicogenic substances in panic patients simply reflects differences in baseline apprehension rather than a direct exacerbation of the neurobiologic system postulated to underlie the disorder. However, the data endorsing such a view has been variable. For instance, whereas some investigators have reported that anticipatory autonomic reactivity prior to challenge administration enhances subsequent panicogenic response to the challenge in panic patients,[40,41] others have reported data suggesting that the behavioral consequences of panicogens are, in large part, independent of baseline level of anxiety.[42,43]

If sensitivity to biological challenges is merely a matter of arousal threshold, it follows that CCK-4 should trigger panic in any highly aroused subject. In fact, data derived from our laboratory suggest that baseline anxiety has little effect on subsequent panicogenic response to CCK-4. For instance in healthy volunteers, prechallenge anxiety, autonomic arousal and neuroendocrine activity were found to be comparable for panickers and non-panickers.[44] In panic patients, response to placebo injection is neither substantial nor significantly different from that observed in healthy volunteers.[5,45] Moreover, in panic patients who had been previously challenged with CCK-4, subsequent challenge was associated with greater increases in baseline anxiety but not greater reactivity to CCK-4.[46] Taken together, these data provisionally suggest that anticipatory response to systemic CCK-4 does not increase the propensity to panic in either panic patients or healthy volunteers. Nevertheless, it is important to note that anticipation of stressful encounters does have a profound effect on CCK release.[46] Moreover, Harro et al[47] have postulated that CCK is a normal endogenous anticipatory stress modulator that may contribute to the genesis of anxiety disorders due to some alteration in the normal regulation of CCK. Conceivably, conspicuous cortical CCK receptor density is a correlate of either anticipatory anxiety and/or deficits in behavioral adaptation to stress.

## SUMMARY

In summary, it appears that psychological factors do not play a particularly crucial role in the induction of panic anxiety states by systemic CCK-4 administration. Evaluation of cognitive appraisal processes, anxiety sensitivity and baseline anxiety revealed that these variables do not fully account for CCK-4 panicogenesis. It should be remembered that the data presented herein are preliminary and further empirical investigations are needed to obtain a clearer understanding of the relationship between psychological factors and panicogenic response to CCK-4.

## REFERENCES


1. Nutt D, Lawson C. Panic attacks: A neurochemical overview of models and mechanisms. Br J Psychiatry 1992; 160:165-178.
2. Clark DM. Cognitive mediation of panic attacks induced by biological challenge tests. Adv Behav Res Therapy. 1993; 15:75-84.
3. Rapee RM. Psychological factors influencing the affective response to biological challenge procedures in panic disorder. J Anx Disorders 1985; 9:59-74.
4. Bradwejn J, Koszycki D, Meterissian G. Cholecystokinin tetrapeptide induces panic attacks in patients with panic disorder. Can J Psychiatry 1990; 35:83-85.
5. Bradwejn J, Koszycki D, Shriqui C. Enhanced sensitivity to cholecystokinin tetrapeptide in panic disorder: Clinical and behavioral findings. Arch Gen Psychiatry 1991; 48:603-610.
6. Lydiard RB, Ballenger JC, Laraia MT et al. CSF cholecystokinin concentrations in patients with panic disorder and in normal comparison subjects. Am J Psychiatry 1992; 149:691-693.
7. Brambilla F, Bellodi L, Perna G et al. Lymphocyte cholecystokinin concentrations in panic disorder. Am J Psychiatry 1993; 150:1111-1113.
8. Clark DM. A cognitive approach of panic. Behav Res Ther 1986; 24:461-470.
9. Mellman TA, Uhde TW. Sleep panic attacks: New clinical findings and theoretical implications. Am J Psychiatry 1989; 146:1204-1207.
10. Mellman TA, Uhde TW. Electroencephalographic sleep in panic disorder: A focus on sleep-related panic attacks. Arch Gen Psychiatry 1989; 46:178-184.
11. Heide F, Borkovec T. Relaxation-induced anxiety: Paradoxical anxiety enhancement due to relaxation training. J Consult Clin Psychol 1983; 51:171-182.
12. Rachman S, Levitt K, Lopatka C. Panic: The links between cognitions and bodily symptoms-I. Behav Res Ther 1987; 25:411-423.
13. Rachman S, Lopatka C, Levitt K. Experimental analyses of panic-II. Panic patients. Behav Res Ther 1988; 26:33-40.
14. Rapee R, Mattick R, Murrell E. Cognitive mediation in the affective component of spontaneous panic attacks. J Behav Ther Exper Psychiatry 1986; 17:245-253.
15. Clark DM, Salkovskis PM, Anastasiades P. Cognitive mediation of lactate induced panic. Paper presented at the 24th annual AABT convention, San Francisco (Abstract).

16. Papp LA, Martinez J, Klein DF et al. 1993. Instructional set and panic disorder. New Research Program and Abstracts from the 146th Annual American Psychiatry Association Meeting. San Francisco, CA.
17. Koenigsberg, HW, Pollak CP, Fine J et al. Cardiac and respiratory activity in panic disorder: Effects of sleep and sleep lactate infusions. Am J Psychiatry 1994; 151:1148-1152.
18. Den Boer JA, Westenberg HGM. Behavioral, neuroendocrine, and biochemical effects of 5-hydroxytyptophan administration in panic disorder. Psychiatry Res 1990; 31:267-278.
19. Koszycki D, Cox BJ, Bradwejn J. Anxiety sensitivity and response to CCK-4 in healthy volunteers. Am J Psychiatry 1993; 150:1881-1883.
20. Koszycki D, Zacharko R, Bradwejn J. The influence of personality on behavioral response to CCK-4 challenge in patients with panic disorder. Submitted.
21. Koszycki D, Bradwejn J. Cognitive and noncognitive panic attack. In preparation.
22. Reiss S, Peterson R, Gorsky D et al. Anxiety sensitivity, anxiety frequency and the prediction of fearfulness. Behav Res Ther 1986; 24: 1-8.
23. Maller RA, Reiss S. Anxiety sensitivity in 1984 and panic attacks in 1987. J Anx Disorders 1992; 6:241-247.
24. Donnell CD, McNally RJ. Anxiety sensitivity and panic attacks in a nonclinical population. Behav Res Ther 1990; 28:83-85.
25. Stewart SH, Knize K, Pihl RO. Anxiety sensitivity and dependency in clinical and non-clinical panickers and controls. J Anx Disorders, 1992; 6:119-131.
26. Holloway W, McNally RJ. Effects of anxiety sensitivity on the response to hyperventilation. J Abnor Psychology 1987; 96:330-334.
27. Rapee RM, Brown TA, Antony MM et al. Response to hyperventilation and inhalation of 5.5% carbon-dioxide enriched air across the DSM-III-R anxiety disorders. J Abnor Psychology 1992; 101:538-552.
28. Rapee RM, Medoro L. Fear of physical sensations and trait anxiety as mediators of the response to hyperventilation in nonclinical subjects. J Abnor Psychology 1994; 103:693-699.
29. Koszycki D. Anxiety sensitivity and panic attacks provoked by CCK-4 in healthy volunteers. In preparation.
30. Price H, Goddard AW, Barr LC et al. Pharmacological challenges in anxiety disorders. In, Bloom FE and Kipfer DJ (Eds.) Psychopharmacology: The fourth generation of progress, Raven Press, New York, 1995:1331-1325.
31. Bradwejn J, Koszycki D. The cholecystokinin hypothesis of anxiety and panic disorders. In, Reeve, J.R., Eysselein, V., Solomon, T.E. & Go, V.L.W. (Eds.), Ann NY Acad Sci, 1994; 173: 273-282.
32. Hathaway SR, McKinley JC. The Minnesota Multiphasic Personality Inventory manual. New York, Psychological Corporation. 1983.

33. Welsh GS. MMPI profiles and factors A and R. J Clin Psychology 1965; 21:43-47.
34. Uvnäs-Moberg K, Arn I, Theorell T et al. Personality traits in a group of individuals with functional disorders of the gastrointestinal tract and their correlation with gastrin, somatostatin and oxytocin levels. J Psychosomatic Res 1991; 35:515-523.
35. Uvnäs-Moberg K, Arn I, Jonsson C-O et al. The relationships between personality traits and plasma gastrin, cholecystokinin, somatostatin, insulin, and oxytocin levels in healthy women. J Psychosomatic Res 1993; 37:581-588.
36. Vasar E, Peuranen E, Harro J et al. Social isolation of rats increases the density of cholecystokinin receptors in the frontal cortex and abolishes the anti-exploratory effect of caerulein. Naunyn-Schmiedeberg's Arch Pharmacol 1993; 348:96-101.
37. Gray JA. The neuropsychology of anxiety: An enquiry into the functions of the septo-hippocampal system. Oxford University Press, Oxford. 1982.
38. Eysenck HJ. The biological basis of personality. Springfield, Ill.: Thomas. 1967.
39. Margraf J, Ehlers A, Roth WT. Biological models of panic disorder and agoraphobia: A review. Behav Res Ther 1986; 24:553-567.
40. Den Boer JA, Westenberg HGM, Klompmakers AA et al. Behavioral, biochemical and neuroendocrine concomitants of lactate-induced panic anxiety. Biol Research 1989; 26:612-622.
41. Liebowitz MR, Fyer AJ, Gorman JM et al. Lactate provocation of panic attacks: K. Clinical and behavioral findings. Arch Gen Psychiatry 1984; 41:764-770.
42. Griez E, Zandbergen J, Pols H et al. Response to 35% $CO_2$ as a marker of panic in severe anxiety. Am J Psychiatry 1990; 147:796-797.
43. Perna G, Battaglia M, Garberi A et al. Carbon dioxide/oxygen challenge test in panic disorder. Psychiatry Res 1994; 52:159-171.
44. Koszycki D, Zacharko RM, Le Melledo J-M et al. The effect of CCK-4 in normal individuals: Comparison of panickers and nonpanickers. Annual Meeting of the Canadian College of Neuropsychopharmacology, Vancouver, Canada, June 5, 1995 (Abstract). 1995.
45. Bradwejn J, Koszycki D, Couetoux du Tertre A et al. The panicogenic effects of cholecystokinin tetrapeptide are antagonized by L-365,260, a central cholecystokinin receptor antagonists, in patients with panic disorder. Arch Gen Psychiatry 1994; 51:486-493.
46. Phillip E, Wilckens T, Friess E et al. Cholecystokinin, gastrin and stress hormone response in marathon runners. Peptides 1992; 13:125-128.
47. Harro J, Vasar E, Bradwejn J. CCK in animal and humans research in anxiety. Trends Pharmacol Sci 1993; 141:244-249.

CHAPTER 6

# Interactions Between Cholecystokinin and Other Neurotransmitter Systems

Jacqueline N. Crawley

## INTRODUCTION

Panic induced by cholecystokinin tetrapeptide in humans is a robust, well-replicated and fascinating phenomenon. Panic-like behavioral and physiological symptoms induced by CCK-4 or pentagastrin have been documented by several major laboratories.[1-10] Careful, double-blind, placebo-controlled studies have demonstrated a consistent, dose-related action of CCK-4. The symptoms expressed are similar for panic patients and normal controls.[2,4] Patients suffering from panic disorder are sensitive to lower doses of CCK-4 than are normal controls. The symptoms expressed are similar for panic patients and normal controls.[3-5,11]

Interestingly, CCK-4-induced panic is a phenomenon in search of a mechanism. This chapter speculates on possible sites of action of cholecystokinin tetrapeptide (CCK-4), in the brain and in the periphery. Relevant literature is described on the interactions of cholecystokinin with other neurotransmitter systems. Neurochemical and anatomical hypotheses are suggested. Experiments to test these hypotheses are proposed.

*Cholecystokinin and Anxiety: from Neuron to Behavior,* edited by Jacques Bradwejn and Eero Vasar. 

## CENTRAL NERVOUS SYSTEM

Cholecystokinin and its two receptor subtypes are widely distributed in the mammalian central nervous system.[12-14] CCK-A and CCK-B receptor subtypes are differentiated by: a) their non-homologous cDNA sequences,[15,16] b) the 1,000-fold selectivity of the CCK-A subtype for sulfated CCK octapeptide, c) the relative nonselectivity of the CCK-B subtype for the unsulfated octapeptide and for CCK-4 and d) the growing number of synthetic agonists and antagonists which are highly selective for one CCK receptor subtype (reviewed in ref. 17).

## CCK/GABA

The highest concentrations of cholecystokinin (CCK) are localized in the cerebral cortex.[18] In the human brain, CCK-like immunoreactivity is present in concentrations of 300-1,200 pmoles per gram wet weight of neocortex.[19] The highest levels of CCK messenger RNA in human brain are detected in the neocortex, indicating that CCK expressed in cortex is present in cell bodies.[20] In the rat, cortical CCK-like immunoreactive cells are non-pyramidal neurons, abundant in cortical layers II and III, frequently bipolar but also multipolar.[12,21] In cats and monkeys, almost all of the CCK-like immunoreactive neurons in the cortex are also immunoreactive for glutamic acid decarboxylase, the synthetic enzyme for the inhibitory neurotransmitter, gamma-aminobutyric acid (GABA).[22,23] This coexistence of CCK with GABA raises the possibility that CCK may interact with GABA in cortical regions relevant to panic-related behaviors.

Benzodiazepines, the major class of anti-anxiety drugs, act at a modulatory site on the GABA receptor complex.[24] Early neurophysiological studies of CCK revealed that benzodiazepines reduced the excitatory actions of CCK in rat brain.[25] Chronic treatment with diazepam or flurazepam similarly reduced the excitatory effect of microiontophorectically applied CCK.[26] However, there is no evidence to date that benzodiazepines displace CCK binding, or that CCK or CCK antagonists displace benzodiazepine binding.[27] Thus, CCK and benzodiazepines bind to unrelated specific high affinity receptors, and do not appear to interact directly through each others' receptors. One hypothesis, therefore, is that CCK-4 produces an excitatory postsynaptic effect which blocks an inhibitory postsynaptic effect of GABA, downstream from the cell

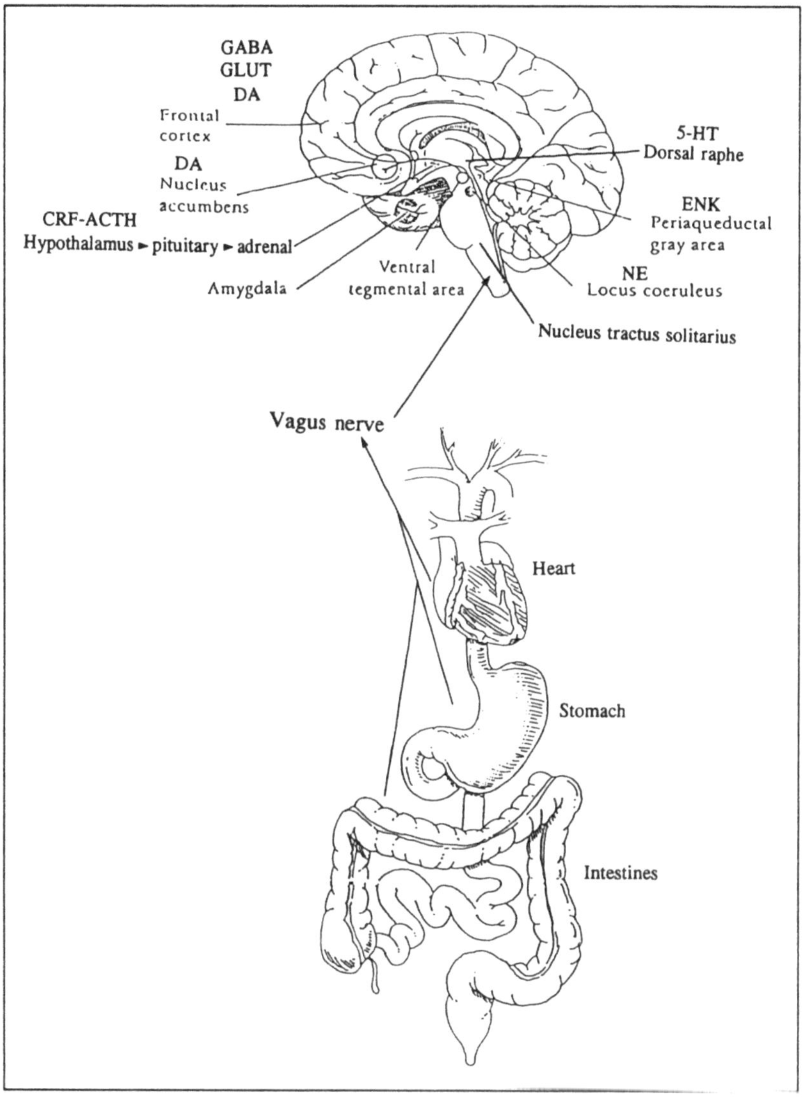


*Fig. 6.1. Possible sites at which CCK-4 may induce panic-like symptoms in human normal volunteers and in patients with panic disorder. 1) frontal cortex, via interactions with gamma aminobutyric acid (GABA), glutamate (GLUT), dopamine (DA); 2) mesocorticolimbic pathway, i.e., dopaminergic pathways from ventral tegmentum to nucleus accumbens and frontal cortex; 3) locus ceruleus via interactions with norepinephrine (NE); 4) dorsal raphe via interactions with serotonin (5-HT); 5) periaqueductal gray and spinal pathways via interactions with endogenous opiates such as enkephalins (ENK); 6) hypothalamic-pituitary-adrenal axis, via interactions with corticotropin releasing factor (CRF), adrenocorticotropin releasing hormone (ACTH), or cortisol; 7) direct activation of brain centers mediating fear responses, e.g., amygdala; direct activation of brain centers mediating respiratory and cardiovascular regulation, e.g., nucleus tractus solitarius; 8) direct activation of the vagus nerve innervating the nucleus tractus solitarius; 9) activation of peripheral organs such as heart, stomach and intestines, through CCK receptors or gastrin receptors, to trigger vagal feedback to the brain.*

bodies in which CCK and GABA coexist. This postulated interaction would occur at a cortical site mediating panic-like behaviors.

Several lines of research have addressed the hypothesis of reciprocal interactions of CCK-4 and GABA. Acute administration of diazepam produced an increase in CCK mRNA levels in rat cortex.[28] Chronic administration of diazepam produced an increase in the number of CCK binding sites in rat olfactory cortex.[29] In rat cerebral cortex slices, CCK increased the potassium-induced release of GABA.[30,31] Behavioral analysis indicated that rats exhibiting higher levels of anxiety-like behaviors on the elevated plus maze had lower densities or numbers of benzodiazepine binding sites than rats exhibiting lower levels of anxiety-like behaviors in this animal model of anxiety.[32] The benzodiazepine antagonist, flumazenil, blocked an anxiety-like action of unsulfated CCK octapeptide on the elevated plus maze in rats, and in the light/dark (black-white) exploration model of anxiety in mice.[33] These several lines of evidence suggest functional interactions between CCK and GABA, although the direction of this interaction is not consistent across paradigms.

The critical experiment has been performed, to test the hypothesis that CCK-4 induces panic-like symptoms in humans through a GABAergic mechanism. The benzodiazepine receptor antagonist, fluamazenil, was administered 15 minutes before an intravenous bolus injection of 50 μg CCK-4, a dose which induces panic in healthy volunteers.[34] Flumazenil was administered at a dose of 2 mg, a dose which has no effect in healthy volunteers but potently blocks the sedative effects of diazepam. Flumazenil had no effect on CCK-4-induced behavioral symptoms of panic, nor on the cardiovascular effects induced by CCK-4.[34] Therefore, while functional interactions between endogenous CCK and GABA may occur at brain pathways relevant to anxiety-related behaviors, the phenomenon of CCK-4-induced panic does not appear to depend upon the benzodiazepine binding site on the GABA receptor.

### CCK/Glutamate

A small number of CCK-like immunoreactive pyramidal-shaped neurons were identified in the rat cerebral cortex.[35] Retrograde tracing analysis revealed that these CCK-containing neurons project to the striatum.[35] The well-known neurotransmitter in the major corticostriatal pathway of the rat is glutamate.[36] Speculations have been raised that CCK coexists with glutamate in the corticostriatal pathway.[35]

A few functional studies of CCK interactions with glutamate, in this and other pathways, have been conducted. CCK increased basal, but not potassium-stimulated, release of glutamate in rat hippocampal slices.[37,38] CCK-4 was reported to exacerbate an ischemia-induced decrease in the CA1 presynaptic potential, presumably mediated by glutamatergic Schaffer collaterals, in rat hippocampal slices.[39] In contrast, a neuroprotective effect of CCK on glutamate-induced neuronal cell death has been reported.[40]

The hypothesis concerning CCK-glutamate interactions as a mechanism for CCK-4-induced panic would postulate that CCK-4 activates a glutamatergic pathway which mediates panic symptoms. This hypothesis could be tested directly in animal models in which CCK-4 produced anxiety-related behaviors and cardiovascular effects. If selective glutamate receptor subtype antagonists blocked CCK-4-induced behavioral and physiological actions, then the anatomical site(s) mediating glutamate/CCK-4 interactions could be analyzed through site-specific microinjections.

### CCK/Dopamine

CCK coexists with dopamine (DA) in ventral tegmental neurons projecting to mesocortical regions in the rat brain.[41,42] One of the terminal fields of this coexsitence is the medial posterior nucleus accumbens, which coincides with the "shell" region of the nucleus accumbens.[41] The shell has a unique set of afferents and efferents that distinguishes it from the core and rostral pole of the nucleus accumbens, and from the caudate nucleus.[43] Further, the distribution of the D-3 dopamine receptor subtype is uniquely localized in the same shell region as the terminal field of the CCK-DA coexistence in the rat.[44]

The mesocorticolimbic dopamine system has been implicated in reward, approach and appetitive behaviors, self-stimulation, self-administration of drugs of abuse and in schizophrenia.[45] Investigations of the actions of cholecystokinin in the mesolimbic pathway of this coexistence have shown that CCK acts as a modulator of dopaminergic functions in the rat.[46-51] When microinjected into the medial posterior accumbens, CCK potentiates dopamine-stimulated adenylate cyclase,[52] potentiates dopamine-induced and amphetamine-induced hyperlocomotion[47,49-51,53] and potentiates self-stimulation of the ventral tegmentum.[54,55] The CCK-A receptor subtype appears to mediate these modulatory actions of CCK in the shell region.[47,51] In the rostral pole of the nucleus accumbens, where CCK is present but does not coexist with DA,[41] CCK inhibits dopamine-stimulated adenylate cyclase, and inhibits dopamine-induced and amphetamine-induced hyperlocomotion.[47,50,52,53] The CCK-B receptor subtype appears to mediate these inhibitory actions of CCK in the rostral pole.[47] In addition, CCK abolished a cyclic AMP-dependent phosphorylation of a dopamine- and cAMP-regulated phosphoprotein, DARPP-32, in the striatum, through a CCK-B receptor subtype mechanism.[56] Therefore, CCK appears to act as a modulator of dopaminergic functions, potentiating the actions of DA in the shell region of coexistence, and inhibiting the actions of DA in the rostral pole region. Further, at the ventral tegmental and substantia nigra cell bodies, CCK has excitatory actions alone, while potentiating the inhibitory actions of DA on neuronal firing rates.[57,58] Acute administration of CCK-B antagonists dose-dependently decreases the number of spontaneously active ventral tegmental neurons.[59] Chronic administration of CCK-B antagonists also inhibits A9 and A10 neuronal activity, an effect which may be mediated through feedback pathways from forebrain sites.[60] These findings of CCK-DA interactions in the mesolimbic pathway have been interpreted in terms of the role of CCK as a modulator of dopaminergic functions relevant to schizophrenia and to drug abuse,[47,48] rather than as a mechanism for CCK-4-induced panic.

More relevant to CCK-4-induced panic may be the mesocortical pathway, which has been strongly implicated in stress responses.[61] Exposure to mild stressors selectively increases DA release in the prefrontal cortex (PFC).[61,62] To increase DA release in the nucleus accumbens or caudate nucleus, higher intensities or longer durations of a stressor are required.[61,63] Conditioned stress activates dopamine neurons in the ventral tegmental area but not in the substantia nigra.[64] The mesocortical dopamine pathway, and to a lesser extent the mesolimbic dopamine pathway, therefore, represent brain pathways responding to stressors. Further, mesocortical dopamine terminals in the prefrontal cortex act on the glutamatergic corticostriatal pathway, to mediate the release of dopamine in the caudate nucleus and nucleus accumbens.[65] In support of this hypothesized circuitry, depletion of prefrontal cortical DA enhanced the responsiveness of nucleus accumbens DA innervation to stress.[66] This trans-synaptic regulation of subcortical DA systems by the mesocortial cortical DA system[65] could occur through the glutamatergic/CCK projections from the cortex to the striatum, as described above.

Exogenously administered CCK is exitatory when microiontophoresed onto midbrain dopaminergic neurons, including ventral tegmental neurons projecting to PFC of the rat.[57] CCK antagonists decrease the number of spontaneously active ventral tegmental neurons, possibly by a feedback pathway originating in the cortex of the rat.[59,60] By activating mesocortical neurons, exogenously administered CCK-4 could contribute to stress responses during a panic attack.

The hypothesis that CCK-4 induces panic attacks by activating mesocortical neurons could be tested using animal models of panic and anxiety. CCK-4 administered into the ventral tegmentum and/or prefrontal cortex would be predicted to induce panic-like and/or anxiety-like behaviors in rodent and primate models. While a large number of behavioral studies have been performed to date with CCK-4 and CCK-B antagonists, these experiments with animal models of anxiety employed systemic or intraventricular routes of administration. Direct microinjection of CCK-4 into regions of the mesocortical dopamine pathway, in sensitive animal models relevant to human panic responses, would be a straightforward method to test this hypothesis.

### CCK/SEROTONIN

Panic disorders are presently treated most effectively with chronic administration of antidepressants, acting at the serotonin transporter or as monoamine oxidase inhibitors, or with agonists acting at the benzodiazepine/GABA receptor complex.[67] In particular, serotonergic pathways in the brain have been implicated in panic disorders and generalized anxiety-related behaviors.[68-70] Serotonergic receptor antagonists have been reported to produce anxiolytic actions in rodent paradigms.[69] Increases in extracellular concentrations of serotonin have been reported in the prefrontal cortex during anxiety-like behaviors in the guinea pig.[71]

Investigations of cholecystokinin-serotonin interactions have revealed excitatory effects of CCK on dorsal raphe serotonergic neurons, primarily mediated by a CCK-A receptor subtype.[72] Treatment with the CCK-B agonist, BOC-CCK-4, potentiated the rise in serotonin in the guinea pig cerebral cortex, during performance of the elevated plus maze task, an animal model of anxiety-related behaviors.[73] Pretreatment with the CCK-B antagonist, L-365,260, blocked the anxiety-related behavioral response on the plus maze task, and prevented the rise in serotonin in cerebral cortex during performance of the task.[73] In addition, 5-$HT_3$ receptor antagonists augmented the release of cholecystokinin from rat frontal cortex,[74] suggesting a reciprocal interaction between CCK and serotonin.

These findings suggest that CCK potentiates the anxiety-related actions of serotonin, and that CCK antagonists may block the anxiety-related actions of serotonin, through a site on dorsal raphe neurons. If endogenous panic attacks are normally initiated by a serotonergic mechanism, then administration of CCK-4 could induce panic through activation of the same serotonergic system. This hypothesis could be tested by pretreatment with a serotonin receptor antagonist, which would be predicted to prevent CCK-4-induced panic in patients or normal volunteers. While this experiment has not been done, a related approach has been taken.[75] Panic patients were treated with chronic imipramine, a serotonin reuptake blocker and standard antidepressant, and then tested with the CCK-4 challenge dose of 20 μg. Imipramine pretreatment significantly reduced the panic symptoms induced by CCK-4, as compared to the panic symptoms shown by the same patients treated with the same dose of CCK-4 before the beginning of imipramine pretreatment.[75] This pilot study indicated that treatments with imipramine which were effective in preventing spontaneous panic attacks were also effective in preventing CCK-4-induced panic.[75] Chronic imipramine can affect many other neurotransmitter systems through long-term adaptations of complex circuitry. Therefore, it is possible that the ability of imipramine to block CCK-4-induced panic is due to the action of imipramine on several systems. However, the findings of this study are consistent with the notion of CCK-serotonin interactions as a mechanism of action of CCK-4 in producing panic symptoms.

### CCK/Norepinephrine

The noradrenergic projections of the locus ceruleus have been implicated in the detection of novel stimuli, in vigilance and anxiety-related behaviors and in stress responses.[76-78] An interesting question is whether CCK-4 activates locus ceruleus neurons, which contribute to the panic response. Experiments on the effects of CCK and CCK-4 on the firing rate of locus ceruleus neurons would be important to address the hypothesis of CCK-NE interactions relevant to CCK-4-induced panic.

One study investigated the effects of locus ceruleus lesions, using the neurotoxin DSP-4, on cholecystokinin receptor binding.[79] [$^3$H]p-CCK-8 binding was reported to be higher in frontal cortex and hippocampus of rats treated with DSP-4, 50 mg/kg i.p., as compared to saline-treated controls.[79] This result could be interpreted as indicating that reduced noradrenergic input increases sensitivity to CCK.

### CCK/Opiates

CCK interacts with opiates on a variety of analgesia paradigms. In particular, the CCK-B antagonists, L-365,260 and CI-988, enhance the analgesic effects of morphine, and may prevent the development of tolerance to morphine.[80-84] CCK has been reported to produce nociceptive and antinociceptive effects, depending on the dose, receptor subtype and route of administration.[85-87] It is interesting to speculate that CCK-4 administration could activate a nociceptive pathway, e.g., by blocking endogenous opiate systems in peripheral organs, spinal cord and/or brain sites. Interoceptive stimuli generated by a decrease in opiate tone could then produce further internal cues, which are interpreted by the patient in a manner which leads to panic symptoms. This hypothesis could be directly tested by pretreatment with an opiate agonist, to increase opiate tone, which would be predicted to prevent CCK-4-induced panic. Alternatively, it would be interesting to test whether naloxone, a standard opiate receptor antagonist, produces panic-like symptoms in panic patients and in normal controls.

## PERIPHERY

### Hypothalamus/Pituitary/Adrenal Axis

Systemically administered CCK produces a large and immediate increase in plasma levels of adrenocorticotrophin hormone (ACTH) and corticosterone in awake rats[88,89] and in man.[90] This effect appears to be mediated by corticotropin releasing factor (CRF), since pretreatment with a CRF antagonist or with CRF antiserum blocked the CCK-induced increase in corticosterone.[89] CCK activates the expression of the early immediate gene, c-fos, in hypothalamic neurons which contain CRF.[91] CCK stimulated the release of CRF in hypothalamic organ culture in vitro[88] and CCK stimulated the release of ACTH from anterior pituitary cells in vitro.[92] These findings suggest that systemically administered CCK can act at circumventricular sites in the hypothalamus, activating the hypothalamic-pituitary-adrenal axis. CCK-4 could be activating the HPA axis, inducing interoceptive cues which could lead to a panic response. However, while activation of the HPA axis is associated with a variety of stress responses, patients with panic disorder show a blunted ACTH response to CRF treatment.[9]

A first test of the hypothesis that CCK-4 induces panic by activating the HPA axis would be to determine whether CCK-4 administration to humans produces a rise in plasma ACTH and cortisol. Studies to date indicate that CCK-4, pentagastrin and cerulein produce increases in plasma cortisol and ACTH levels.[1,9,90] A second test of this hypothesis would be to determine whether treatments which block the HPA response to stressors, such as a CRF antagonist, would prevent CCK-4-induced panic responses.

### Nucleus Tractus Solitarius/Vagus Nerve

The nucleus tractus solitarius (NTS) is located in the hindbrain, adjacent to the fourth ventricle. CCK receptors, of both the CCK-A and the CCK-B subtypes, are localized on NTS cell bodies.[13,14] Peripheral administration of sulfated CCK octapeptide induces expression of the immediate early gene, c-fos, in the NTS, in the adjacent area postrema and in forebrain structures.[93-97] Neurophysiological analysis indicates that activation of CCK-B receptors on the NTS, by the selective CCK-B agonist, BC 264, produces prolonged excitation of NTS neuronal firing, while activation of CCK-A receptors on the NTS produces delayed inhibition of NTS neuronal firing.[98] Systemically administered CCK-4 may reach CCK-B receptors on NTS neurons through the capillary network in the ventricular system, which bathes circumventricular structures such as the NTS. NTS neurons activated by CCK-4 could then activate forebrain structures which contribute to panic-related symptoms. Relevant forebrain structures innervated by the NTS include the supraoptic, medial preoptic, arcuate, dorsomedial and paraventricular nuclei of the hypothalamus, periventricular nucleus of the thalamus, central nucleus of the amygdala and bed nucleus of the stria terminalis.[99,100]

Systemically administered CCK-4 may activate peripheral mechanisms which then send signals to central sites. The NTS is innervated by afferents from the amygdala, and from cranial nerves VII, IX and X.[101,102] In particular, CCK binding sites have been detected in the vagus nerve.[103-107] CCK-A receptors are localized to vagal afferent fibers projecting principally to the medial NTS; CCK-B receptors are localized to vagal afferent fibers projecting principally to the lateral NTS, in the rat.[108] In the rabbit, subpopulations of neurons in the vagal afferent nodose ganglion express CCK-A and CCK-B receptors.[109] Lesion of the vagal afferents with capsaicin prevents CCK-induced expression of c-fos in the NTS and area postrema.[110] Blockade of CCK-A receptors, but not of CCK-B receptors, attenuates CCK-induced satiety.[111] These findings indicate that sulfated CCK octapeptide, administered in the periphery, may activate CCK-A receptors in the vagus nerve, which then activate the medial NTS neurons, which then project to the forebrain regions involved in the inhibitory regulation of feeding behavior.[112-114]

In contrast to vagal CCK-A receptors involved in satiety, CCK-B receptors in the vagus nerve and NTS may mediate excitatory actions relevant to panic-like symptoms. Vagal afferent fibers to the NTS arise from several peripheral organs which may be relevant to panic-like symptoms. The stomach is a primary source of afferent vagal fibers terminating in the medial NTS.[115] In addition, hemodynamic actions of CCK may be relayed through the vagus nerve. Peripherally administered CCK caused an increase in heart rate and blood pressure in the cat,[116] and a brief bradycardia, followed by prolonged tachycardia in rats.[117] Blood pressure increased concomitant with the bradycardia, followed by mesenteric and renal vasoconstriction.[117] CCK may act directly on respiratory centers in the NTS, as short-term decreases in respiration are observed after CCK administration in cats.[118] In addition, a central CCK-B receptor appears to mediate the fever-like thermoregulatory response induced by the CCK in the rat,[119] while other studies demonstrate a hypothermic action of centrally administered CCK.[120,121]

Patients with panic disorder show an increase in heart rate and diastolic blood pressure after CCK-4 administration.[4] It is interesting to speculate that changes in heart rate, blood pressure or body temperature could be interpreted by a panic disorder patient, or by a normal volunteer, as disturbing interoceptive cues which then precipitate panic-like symptoms. Does CCK-4, administered as an iv bolus to humans, have effects on heart rate, blood pressure and body temperature, via CCK-B receptors in the NTS and/or vagus nerve, similar to those reported with sulfated CCK octapeptide in animal experiments? Experiments with CCK-4 could be performed in the same elegant animal physiology paradigms previously used to test sulfated CCK octapeptide, to answer this critical question about the actions of CCK-4 on CCK-B receptors in the vagus nerve.

### Gastrointestinal Tract and Gastrin Receptors

The cDNA sequence for the gastrin receptor appears to be identical with the CCK-B receptor, as elucidated from rat brain, rat pancreas and canine parietal cells.[15,16,122] This fascinating discovery raises the possibility that CCK-4 acts directly on gastrin receptors in the stomach. The effects of CCK on gastric emptying,[123,124] relevant to inhibition of feeding,[125] appear to be mediated by a CCK-A mechanism.[126]

In the stomach, however, CCK stimulates gastric acid secretion in parietal cells[127] via a CCK-B/gastrin receptor mechanism.[128] CCK also inhibits gastric acid secretion through the release of somatostatin from cells in the gastric mucosa,[129] however this effect appears to be mediated by a CCK-A receptor subtype.[130,131] It is interesting to speculate that an intravenous bolus injection of CCK-4 acts on gastrin receptors in the stomach, to produce a significant elevation in gastric acid secretion which could be interpreted as the fabled "butterflies in the stomach" associated with fear reactions. Consistent with this hypothesis are self-reports from some social phobic patients and healthy volunteers that pentagastrin produces epigastric discomfort and burning sensations in the lower esophageal/stomach region, analogous to gastric acid reflux (Una McCann, National Institute of Mental Health, personal communication).

Further, CCK may act directly on the intestinal tract. Intravenous bolus administration of both CCK and CCK-4 induced nausea, gastrointestinal cramping, and abdominal distress in normal volunteers, although these symptoms were much milder and of shorter duration after CCK-4 than after CCK.[1,3] A role for cholecystokinin sensitivity in irritable bowel syndrome, linked to some anxiety disorders, has been proposed.[132,133] An association between endogenous levels of gastrin and anxiety-like personality traits has been suggested.[134] It will be important to assay blood levels of CCK-4, versus blood levels of gastrin, under baseline conditions in panic patients. The critical, but difficult, experiment would be to assay blood levels of CCK-4 and gastrin during spontaneous panic attacks, in panic patients. Such assays could determine whether elevated levels of these agonists of the CCK-B receptor are temporally correlated with symptoms in panic patients. Further approaches to test the role of gastric and intestinal actions of CCK-4 in panic symptomology could involve pretreatment with anti-emetic drugs that block nausea in humans, and pretreatment with gastric acid inhibiting drugs that block gastric acid secretion.

## ISSUES OUTSTANDING

Several major experimental problems remain to be solved, before it will be possible to definitively determine the mechanism by which CCK-4 induces panic symptoms.

### BLOOD-BRAIN BARRIER

CCK-4 induces panic in humans when administered as a bolus intravenous injection. Does this CCK-4 reach the brain, in quantities sufficient to produce direct central effects? Most peptides are quickly metabolized by peptidases in the liver, bloodstream and brain capillary epithelial cells.[135] A bolus injection may introduce sufficient quantities of CCK-4 to quickly reach the carotid artery, and enter the brain, before the first pass through the liver. This hypothesis could be tested by administering a bolus injection of radiolabelled CCK-4 iv and measuring the authentic radiolabelled CCK-4 in the brain, at time points corresponding to the time course of the panic symptoms. In mice, approximately 0.2% of the selective CCK-B ligand, [$^3$H]-pBC 264, was detected in the brain after intraperitoneal administration.[136] Analogous studies are needed in human imaging paradigms.

### SPECIFICITY FOR THE CCK-B RECEPTOR SUBTYPE

A proper study to compare CCK-A versus CCK-B agonists and antagonists would require full dose response curves for the comparable compounds. Unfortunately, CCK-8 induces considerably more gastrointestinal discomfort than CCK-4, at comparable doses.[1] Does this mean that higher doses of CCK-8 cannot be tested properly to quantitate the incidence of panic-like symptoms, because of the prevalence of gastrointestinal distress? If so, specificity of CCK-4-induced panic for the CCK-B receptors can be questioned. Full dose-response curves, for agonists and antagonists selective for CCK-A versus CCK-B receptors, are necessary to conclusively determine receptor subtype specificity. This approach may not be feasible in humans because of the issue of abdominal distress.

## Animal Models

CCK-4 has been reported to induce anxiety-like behaviors, and CCK-B antagonists have been reported to induce anxiolytic-like actions, in rats, mice and monkeys.[33,137-153] These experiments have employed established animal models of anxiety-like behaviors, including the elevated plus maze, the light ↔ dark exploratory transitions paradigm, open field, olfactory recognition, social recognition, separation-induced ultrasonic vocalizations, predator-induced freezing, the marmoset threat test, restless behaviors in African green monkeys and the Geller-Seifter punished responding conflict test in rats and squirrel monkeys.[32,69,73,137-153] In addition, CCK-4 levels in brain were elevated in several brain regions of rats exposed to the odor of a predator.[148]

Taken together, this large literature on animal models supports the human literature, indicating that CCK-4 induces anxiety-like behaviors through a CCK-B mechanism, and that CCK-B antagonists are anxiolytic. However, taken individually, many of these animal studies show rather small effects of CCK-4 and of CCK-B antagonists, as compared to benzodiazepine antagonists and agonists.[155] In some cases, only one or two doses are significant. The sharp U-shaped dose-response curve may be characteristic of some peptide actions, or may indicate a minor, hard-to-detect effect. In some cases, inappropriate statistical analyses were used, such that the significance of the small effects is questionable. Further, some of the publications which analyzed specificity for the CCK-B subtype did not use full dose-response curves to compare CCK-B versus CCK-A compounds. Some of the publications testing central versus peripheral site of action of CCK-B compounds administered very large doses into the lateral ventricles. Large bolus doses of CCK, given intraventricularly, are likely to leak to the periphery,[114] and may have produced behavioral sedation through a peripheral CCK-A receptor.[156] Behavioral sedation can confound the interpretation of data from many of the behavioral paradigms summarized above. Several negative or contradictory reports have emerged. A careful study by Iversen and co-workers indicated that a CCK-B antagonist did not generalize to the interoceptive discriminative properties of diazepam.[157] In addition, an intravenous pentagastrin challenge failed to induce panic-like symptoms in rhesus monkeys.[158] Our laboratory, which developed the light ↔ dark exploratory transitions model of anxiety-like behaviors, has tested CCK-4 and CCK-B antagonists in this model. We detected no significant effects of these compounds, over a wide dose range, on the number of light ↔ dark transitions or on the amount of time spent in the dark compartment, in C57BL/6J mice that showed significant responses to diazepam in this paradigm (unpublished data). Dose-response curves for the anxiolytic effects of CCK-A and CCK-B antagonists have been found to be similar for the two classes of compounds in one study.[3] Caution must be taken, therefore, against overinterpreting the animal literature. Further, while there are very good animal models for anxiety-related behaviors, which predict anxiolytic drug response in humans, there are presently no clear rodent or primate models specific for panic disorders.

## CONCLUSION

Thoughtful reviews have been published recently, analyzing the phenomenology of CCK-4-induced panic in humans, and the relevant animal literature[159-163] (and see other chapters in this volume). Delineation of the anatomical sites and neuropharmacological mechanisms mediating CCK-4-induced panic will require careful research along several lines. These include analyses of the blood-brain barrier for CCK-4 in humans, assay of endogenous CCK-4 and gastrin levels during panic attacks, development of more appropriate animal models of panic disorder and site-specific microinjections of CCK-4 and CCK-B antagonists in these animal models. It will be intriguing to discover the mechanism responsible for the robust CCK-4 symptomatology. Knowledge of the mechanism for this phenomenon is likely to lead to a better understanding of panic disorder, and the development of more selective and efficacious treatments for this debilitating syndrome.

ACKNOWLEDGMENTS

We thank Dr. Jacques Bradwejn, University of Toronto, Dr. Franco Vaccarino, University of Toronto, Dr. Una McCann, National Institute of Mental Health, and Dr. Itzak Angel, Synthelabo Co., for their stimulating conceptual insights. Ms. Carol Markey provided excellent assistance in the literature search and editing of this chapter.

REFERENCES

1. de Montigny C. Cholecystokinin tetrapeptide induces panic-like attacks in healthy volunteers. Arch Gen Psychiatry 1989; 46:511-517.
2. Bradwejn J, Koszycki, D, Meterissian G. Cholecystokinin-tetrapeptide induces panic attacks in patients with panic disorder. Can J Psychiatry 1990; 35:83-85.
3. Bradwejn J, Koszycki D, Bourin M. Dose ranging study of the effects of cholecystokinin in healthy volunteers. J Psychiatry Neurosci 1991;16:91-95.
4. Bradwejn, J, Koszycki D, Annable, L et al. A dose-ranging study of the behavioral and cardiovascular effects of CCK-tetrapeptide in panic disorder. Biol Psychiatry 1992;32:903-912.
5. Bradwejn J, Koszycki D, Payeur R et al. Replication of action of cholecystokinin tetrapeptide in panic disorder: Clinical and behavioral findings. Am J Psychiatry 1992; 149:962-964.
6. Bradwejn J, Koszycki D, Couetoux du Tertre A et al. The panicogenic effects of cholecystokinin tetrapeptide are antagonized by L-365,260, a central cholecystokinin receptor antagonist, in patients with panic disorder. Arch Gen Psychiatry 1994;51:486-493.
7. Abelson JL, Nesse RM. Cholecystokinin-4 and panic. Arch Gen Psychiatry 1990; 47:395.
8. Abelson JL, Nesse RM. Pentagastrin infusions in patients with panic disorder. I. Symptoms and cardiovascular responses. Biol Psychiatry 1994; 36:73-83.
9. Abelson JL, Nesse RM, Vinik AI. Pentagastrin infusions in patients with panic disorder. II. Neuroendocrinology. Biol Psychiatry 1994; 36:84-96.
10. Uhde TW, Terrill DR, Chambless DL et al. Pentagastrin model of anxiety in humans. American College of Neuropsychopharmacology Abstracts 1993:59.
11. Bradwejn J, Koszycki D, Shriqui C. Enhanced sensitivity to cholecystokinin tetrapeptide in panic disorder. Arch Gen Psychiatry 1991; 48:603-610.
12. Hökfelt T, Cortes R, Schalling M et al. Distribution pattern of CCK and CCK mRNA in some neuronal and non-neuronal tissues. Neuropeptides 1991; 19:31-43.

13. Hill DR, Shaw TM, Graham W et al. Autoradiographical detection of cholecystokinin-A receptors in primate brain using [$^{125}$I]-Bolton Hunter CCK-8 and 3-H-MK-329. J Neurosci 1990; 10:1070-1081.
14. Moran TH, Robinson PH, Goldrich MS et al. Two brain cholecystokinin receptors: implications for behavioral actions. Brain Research 1986; 362:175-179.
15. Wank SA, Pisegna JR, de Weerth A. Brain and gastrointestinal cholecystokinin receptor family: Structure and functional expression. Proc Natl Acad Sci USA 1992; 89:8691-8695.
16. Kopin AS, Lee Y-M, McBride EW et al. Expression cloning and characterization of the canine parietal call gastrin receptor. Proc Natl Acad Sci U S A 1992; 89:3605-3609.
17. Crawley JN, Corwin RL. Biological actions of cholecystokinin. Peptides 1994; 15:731-755.
18. Lindefors N, Linden A, Brene S et al. CCK peptides and mRNA in the human brain. Prog Neurobiol 1993; 40:671-690.
19. Emson PC, Rehfeld JF, Rossor MN. Distribution of cholecystokinin-like peptides in the human brain. J Neurochem 1982; 38:1177-1179.
20. Lindefors N, Brene S, Kopp J et al. Distribution of cholecystokinin mRNA and peptides in the human brain. Neuroscience 1991; 42:813-821.
21. Hendry SHC, Jones EG, Beinfeld MC. Cholecystokinin-immunoreactive neurons in rat and monkey cerebral cortex make symmetric synapses and have intimate associations with blood vessels. Proc Natl Acad Sci U S A 1983; 80:2400-2404.
22. Hendry SHC, Jones EG, DeFillipe J et al. Neuropeptide-containing neurons of the cerebral cortex are also GABAergic. Proc Natl Acad Sci U S A 1984; 81:6526-6530.
23. Somogyi P, Hodgson AJ, Smith AD et al. Different populations of GABAergic neurons in the visual cortex and hippocampus of cat contain somatostatin- or cholecystokinin-immunoreactive material. J Neurosci 1984; 4:2590-2603.
24. Paul SM, Crawley JN, Skolnick P. The neurobiology of anxiety. In: eds Berger P Brodie H. American Handbook of Psychiatry New York, Basic Book, 1987:1-10.
25. Bradwejn J, de Montigny C. Benzodiazepines antagonize cholecystokinin-induced activation of rat hippocampal neurons. Nature 1984; 312:363-364.
26. Bouthillier A, de Montigny C. Long-term benzodiazepine treatment reduces neuronal responsiveness to cholecystokinin: an electrophysiological study in the rat. Eur J Pharmacol 1988; 151:135-138.
27. Chang RSL, Lotti VJ. L-364,718: a review of the biochemical and pharmacological characterization of a highly potent peripherally selective CCK antagonist. In: eds Wang RY, Schoenfeld R. Cholecystokinin Antagonists, Alan R. Liss, Inc., New York, 1988:13-28.

28. Rattray M, Singhvi S, Wu P-Y et al. Benzodiazepines increase preprocholecystokinin messenger RNA levels in rat brain. Eur J Pharmacol 1993; 245:193-196.
29. Harro J, Lang A, Vasar E. Long-term diazepam treatment produces changes in cholecystokinin receptor binding in rat brain. Eur J Pharmacol 1990; 180:77-83.
30. Sheehan M, de Belleroche J. Facilitation of GABA release by cholecystokinin and caerulein in rat cerebral cortex. Neuropeptides 1983; 3:429-434.
31. Perez de la Mora N, Hernandez-Gomez A-M et al. Cholecystokinin-8 increases $K^+$-evoked [$^3$H]gamma-aminobutyric acid release in slices from various brain areas. Eur J Pharmacol 1993; 250:423-430.
32. Harro J, Kiivet R-A, Lang A et al. Rats with anxious or non-anxious type of exploratory behaviour differ in their brain CCK-8 and benzodiazepine receptor characteristics. Behav Brain Res 1990; 39:63-71.
33. Chopin P, Briley M. The benzodiazepine antagonist flumazenil blocks the effects of CCK receptor agonists and antagonists in the elevated plus-maze. Psychopharmacol 1993; 110:409-414.
34. Bradwejn J, Koszycki D, Couetoux du Tertre A et al. Effects of flumazenil on cholecystokinin-tetrapeptide-induced panic symptoms in healthy volunteers. Psychopharmacol 1994; 114:257-261.
35. Morino P, Herrera-Marschitz M, Castel MN et al. Cholecystokinin in cortico-striatal neurons in the rat: immunohistochemical studies at the light and electron microscopical level. Eur J Neurosci 1994; 6:681-692.
36. McGeer PL, McGeer EG, Schere V et al. A glutamatergic corticostriatal pathway. Brain Res 1977; 128:369-373.
37. Migaud M, Roques BP, Durieux C. Effects of cholecystokinin octapeptide and BC 264, a potent and selective CCK-B agonist on aspartate and glutamate release from rat hippocampal slices. Neuropharmacology 1994; 33:737-743.
38. Herrera-Marschitz M, Meana JJ Hökfelt T et al. Cholecystokinin is released from a crossed corticostriatal pathway. NeuroReport 1992; 3:905-908.
39. Minamoto Y, Tanaka T, Shibata S et al. Neuroprotective effect of cholecystokinin-B receptor antagonist on ischemia-induced decrease in CA1 presynaptic fiber spikes in rat hippocampal slices. Neurosci Lett 1994; 167:81-84.
40. Katsuura G, Shinohara S, Shintaku H et al. Protective effect of CCK-8 and ceruletide on glutamate-induced neuronal cell death in rat neuron cultures: possible involvement of CCK-B receptors. Neurosci Lett 1991; 132:159-162.
41. Hökfelt T, Skirboll L, Rehfeld JF et al. A subpopulation of mesencephalic dopamine neurons projecting to limbic areas containing cholecystokinin-like peptides: evidence from immunohistochemistry combined with retrograde tracing. Neuroscience 1980; 5:2093-2124.

42. Seroogy KB, Mehta A, Fallon JH. Neurotensin and cholecystokinin coexistence within neurons of the ventral mesencephalon: projections to forebrain. Exp Brain Res 1989; 74:149-162.
43. Deutch AY, Cameron DS. Pharmacological characterization of dopamine systems in the nucleus accumbens core and shell. Neuroscience 1992; 46:49-56.
44. Levesque D, Diaz J, Pilon C et al. Identification, characterization, and localization of the dopamine $D_3$ receptor in rat brain using 7-[$^3$H]hydroxy-N,N-di-n-propyl-2-aminotetralin. Proc Nat Acad Sci U S A 1992; 89:8155-8159.
45. Kalivas PW, Nemeroff CB. The Mesocorticolimbic Dopamine System. Ann N Y Acad Sci Volume 537, New York, 1988: 540 pages.
46. Crawley JN. Cholecystokinin-dopamine interactions. Trends Pharmacol Sci 1991; 12:232-236.
47. Crawley JN. Subtype-selective cholecystokinin receptor antagonists block cholecystokinin modulation of dopamine-mediated behaviors in the rat mesolimbic pathway. J Neurosci 1992; 12:3380-3391.
48. Crawley JN. Cholecystokinin modulates dopamine-mediated behaviors: Differential actions in medial posterior versus anterior nucleus accumbens. Ann N Y Acad Sci 1994; 713:138-142.
49. Crawley JN, Stivers JA, Blumstein L.K et al. Cholecystokinin potentiates dopamine-mediated behaviors: evidence for modulation specific to a site of coexistence. J Neurosci 1984; 5:1972-1983.
50. Crawley JN, Hommer DW, Skirboll LR. Topographical analysis of nucleus accumbens sites at which cholecystokinin potentiates dopamine-induced hyperlocomotion in the rat. Brain Res 1985; 335:337-341.
51. Crawley JN, Stivers JA, Hommer DW et al. Antagonists of central and peripheral behavioral actions of cholecystokinin. J Pharmacol Exp Ther 1986; 236:320-330.
52. Studler JM, Reibaud M, Herve D et al. Opposite effects of sulfated cholecystokinin on DA-sensitive adenylate cyclase in two areas of the rat nucleus accumbens. Eur J Pharmacol 1986; 126:125-128.
53. Vaccarino FJ, Rankin J. Nucleus accumbens cholecystokinin (CCK) can either attenuate or potentiate amphetamine-induced locomotor activity: Evidence for rostral-caudal differences in accumbens CCK function. Behav Neurosci 1989; 103:831-836.
54. De Witte PC, Heidbreder C, Roques B et al. Opposite effects of cholecystokinin octapeptide (CCK-8) and tetrapeptide (CCK-4) after injection into the caudal part of the nucleus accumbens or into its rostral part and the cerebral ventricles. Neurochem Int 1987; 4:473-479.
55. Vaccarino FJ, Vaccarino AL. Antagonism of cholecystokinin function in the rostral and caudal nucleus accumbens: Differential effects on brain stimulation reward. Neurosci Lett 1989; 97:151-156.
56. Snyder GL, Fisone G, Morino P et al. Regulation by the neuropeptide cholecystokinin (CCK-8S) of protein phosphorylate in the neostriatum. Proc Nat Acad Sci U S A 1993; 90:11277-11281.

57. Skirboll LR, Grace AA, Hommer DW et al. Peptide monoamine coexistence: Studies of actions of cholecystokinin-like peptide on the electrical activity of midbrain dopamine neurons. Neuroscience 1981; 6:2111-2124.
58. Hommer DW, Stoner G, Crawley JN et al. Cholecystokinin-dopamine coexistence: electrophysiological actions corresponding to cholecystokinin receptor subtype. J Neurosci 1986; 6:3039-3043.
59. Rasmussen K, Czachura JF, Stockton JR et al. Electrophysiological effects of diphenylpyrazolidinone CCK-B and CCK-A antagonists on midbrain dopamine neurons. J Pharmacol Exp Ther 1993; 264:480-488.
60. Rasmussen K, Stockton ME, Czachura JF et al. Inhibition of A9 and A10 dopamine cells by cholecystokinin-B antagonist LY262691: Mediation through feedback pathways from forebrain sites. Synapse 1993; 15:95-103.
61. Roth RH, Tam S-Y, Ida Y et al. Stress and the mesocortical dopamine systems. In: ed Kalivas PW, Nemeroff CB. The Mesocorticolimbic Dopamine System, , Ann N Y Acad Sci 1988:537:138-147.
62. Thierry AM, Tassin JP, Blanc G et al. Selective activation of the mesocortical DA system by stress. Nature 1976; 263:242-244.
63. Bannon MJ, Roth RH. Pharmacology of mesocorticolimbic dopamine neurons. Pharmacol Rev 1983; 35:53-68.
64. Deutch AY, Tam S-Y, Roth RH. Footshock and conditioned stress increase 3,4-dihydroxyphenylacetic acid (DOPAC) in the ventral tegmental area but not substantia nigra. Brain Res 1985; 333:143-146.
65. Deutch AY. The regulation of subcortical dopamine systems by the prefrontal cortex: interactions of central dopamine systems and the pathogenesis of schizophrenia. J Neural Transm 1992; 36:61-69, 1992.
66. Deutch AY, Clark WA, Roth RA. Prefrontal cortex dopamine depletion enhances the responsiveness of mesolimbic dopamine neurons to stress. Brain Res 1990; 521:311-315.
67. Klein DF, Liebowitz MR, Gorman JM et al. Evaluation of psychotropic drug treatment of panic-related disorders and social phobia. In: ed Prien RF, Robinson DS. Clinical Evaluation of Psychotropic Drugs: Principles and Guidelines, Raven Press, New York, 1994:411-429.
68. Charney DS, Heninger GR. Serotonergic function in panic disorders. Arch Gen Psychiatry 1986; 43:1059-1065.
69. Chopin P, Briley M. Animal models of anxiety: the effects of compounds that modify 5-HT neurotransmission. Trends Pharmacol Sci 1987; 8:383-388.
70. Hamon M. Neuropharmacology of anxiety: perspectives and prospects. Trends Pharmacol Sci 1994; 15:36-39.
71. Lawrence AJ, Marsden CA. Terminal autoreceptor control of 5-hydroxytryptamine as measured by in vivo microdialysis in the conscious guinea-pig. J Neurochem 1992; 58:142-146.

72. Boden PR, Woodruff GN, Pinnock RD. Pharmacology of a cholecystokinin receptor on 5-hydroxytryptamine neurons in dorsal raphe of the rat. Brit J Pharmacol 1991; 102:635-638.
73. Rex A, Fink H, Marsden CA. Effects of BOC-CCK-4 and L 365,260 and cortical 5-HT release in guinea-pigs on exposure to the elevated plus maze. Neuropharmacology 1994; 33:559-565.
74. Raiteri M, Paudice P, Vallebuona F. Inhibition by 5-$HT_3$ receptor antagonists of release of cholecystokinin-like immunoreactivity from the frontal cortex of freely moving rats. Naunyn-Schmiedeberg's Arch Pharmacol 1993; 347:111-114.
75. Bradwejn J, Koszycki D. Imipramine antagonism of the panicogenic effects of cholecystokinin tetrapeptide in panic disorder patients. Am J Psychiatry 1994; 151:261-263.
76. Aston-Jones G, Rajkowski J, Kubiak P et al. Locus ceruleus neurons in monkey are selectively activated by attended cues in a vigilance task. J Neurosci 1994; 14:4467-4480.
77. Levine ES, Litto WJ, Jacobs BL. Activity of cat locus ceruleus noradrenergic neurons during the defense reaction. Brain Res 1990; 531:189-195.
78. Redmond DE. Studies of the nucleus locus ceruleus in monkeys and hypotheses for neuropsychopharmacology. In: ed Meltzer HY. Psychopharmacology: The Third Generation of Progress, Raven Press, NY, 1987:967-975.
79. Harro J, Jossan SS, Oreland L. Changes in cholecystokinin receptor binding in rat brain after selective damage of locus ceruleus projections by DSP-4 treatment. Naunyn-Schmeideberg's Arch Pharmacol 1992; 346:425-431.
80. Dourish CR, O'Neill MF, Coughlan J et al. The selective CCK-B receptor antagonist L-365,260 enhances morphine analgesia and prevents morphine tolerance in the rat. Eur J Pharmacol 1990; 176:35-44.
81. Wiesenfeld-Hallin Z, Xu X-J, Hughes J et al. PD134308, a selective antagonist of cholecystokinin type B receptor, enhances the analgesic effect of morphine and synergistically interacts with intrathecal galanin to depress spinal nociceptive reflexes. Proc Natl Acad Sci U S A 1990; 87:7105-7109.
82. Woodruff GN, Hughes J. Cholecystokinin antagonists. Ann Rev Pharmacol Toxicol 1991; 31:469-501.
83. Xu X-J, Hökfelt T, Hughes J et al. The CCK-B antagonist CI-988 enhances the reflex-depressive effect of morphine in axotomized rats. NeuroReport 1994; 5:718-720.
84. Maldonado R, Derrien M, Noble F et al. Association of a peptidase inhibitor and a CCK-B antagonist strongly potentiates antinociception mediated by endogenous enkephalins. NeuroReport, 1995 in press.
85. Zetler G. Analgesia and ptosis caused by caerulein and cholecystokinin octapeptide (CCK-8). Neuropharmacology 1980; 19:415-422.

86. Wiertelak EP, Maier SF, Watkins LR. Cholecystokinin antianalgesia: safety cues abolish morphine analgesia. Science 1992; 256:830-833.
87. Derrien M, Noble F, Maldonado R et al. Cholecystokinin-A but not cholecystokinin-B receptor stimulation induces endogenous opioid-dependent antinociceptive effects in the hot plate test in mice. Neurosci Lett 1993; 160:193-196.
88. Kamilaris TC, Johnson EO, Calogero AE et al. Cholecystokinin-octapeptide stimulates hypothalamic-pituitary-adrenal function in rats: Role of corticotropin-releasing hormone. Endocrinology 1992; 130:1764-1774.
89. Biro E, Sarnyai Z, Penke B et al. Role of endogenous corticotropin-releasing factor in mediation of neuroendocrine and behavioral responses to cholecystokinin octapeptide sulfate ester in rats. Neuroendocrinology 1993; 57:340-345.
90. Spath-Schwalbe E, Piroth L, Pietrowsky R et al. Stimulation of the pituitary adrenocortical system in man by cerulein, a cholecystokinin-8-like peptide. Clin Physiol Biochem 1988; 6:316-320.
91. Verbalis JG, Stricker EM, Robinson AG et al. Cholecystokinin activates c-fos expression in hypothalamic oxytocin and corticotropin-releasing hormone neurons. J Neuroendocrinol 1991; 3:205-213.
92. Reisine T, Jensen R. Cholecystokinin-8 stimulates adrenocorticotropin release from anterior pituitary cells. J Pharmacol Experl Ther 1986; 236:621-626.
93. Rinaman L, Verbalis JG, Stricker EM et al. Distribution and neurochemical phenotypes of caudal medullary neurons activated to express C-Fos following peripheral administration of cholecystokinin. J Comp Neurol 1993; 338:475-490.
94. Rinaman L, Hoffman GE, Stricker EM et al. Exogenous cholecystokinin activates C-Fos expression in medullary but not hypothalamic neurons in neonatal rats. Dev Brain Res 1994; 77:140-145.
95. Day HEW, McKnight AT, Poat JA et al. Evidence that cholecystokinin induces immediate early gene expression in the brainstem, hypothalamus, and amygdala of the rat by a $CCK_A$ receptor mechanism. Neuropharmacology 1994; 33:719-727.
96. Chen D-Y, Deutsch JA, Gonzalez MF et al. The induction and suppression of c-fos expression in the rat brain by cholecystokinin and its antagonist L364,718. Neurosci Lett 1993; 149:91-94.
97. Luckman SM. Fos-like immunoreactivity in the brainstem of the rat following peripheral administration of cholecystokinin. J Neuroendocrinol 1992; 4:149-152.
98. Branchereau P, Böhme GA, Champagnat J et al. $Cholecystokinin_A$ and $cholecystokinin_B$ receptors in neurons of the brainstem solitary complex of the rat: pharmacological identification. J Pharmacol Exper Ther 1992; 260:1433-1440.
99. Richardo JA, Koh ET. Anatomical evidence of direct projections from the nucleus of the solitary tract to the hypothalamus, amygdala, and other forebrain structures in the rat. Brain Res 1978; 153:1-26.

100. Sawchenko PE, Swanson LW. Central noradrenergic pathways for the integration of hypothalamic neuroendocrine and autonomic responses. Science 1981; 214:685-687.
101. Schwaber JS, Kapp BS, Higgins GA et al. Amygdaloid and basal forebrain direct connections with the nucleus of the solitary tract and the dorsal motor nucleus. J Neurosci 1982; 2:1424-1438.
102. Palkovits M, Zaborsky L. Neuroanatomy of central cardiovascular control. Nucleus tractus solitarii: afferent and efferent neuronal connections in relation to the baroreceptor reflex arc. Prog Brain Res 1977; 47:9-34.
103. Moran TH, Smith GP, Hostetler AM et al. Transport of cholecystokinin (CCK) binding sites in subdiaphragmatic vagal branches. Brain Res 1987; 415:149-152.
104. Moran TH, Norgren R, Crosby RJ et al. Central and peripheral vagal transport of cholecystokinin binding sites occurs in afferent fibers. Brain Res 1990; 526:95-102.
105. Ladenheim EE, Speth RC, Ritter RC. Reduction of CCK-8 binding in the nucleus of the solitary tract in unilaterally nodosectomized rats. Brain Res 1988; 474:125-129.
106. Lin CW, Miller TR. Both CCK-A and CCK-B/gastrin receptors are present on rabbit vagus nerve. Am J Physiol 1992; 263:R591-R595.
107. Mercer JG, Lawrence CB. Selectivity of cholecystokinin (CCK) receptor antagonists, MK-329 and L-365,260, for axonally-transported CCK binding sites on the rat vagus nerve. Neurosci Lett 1992; 137:229-231.
108. Corp ES, McQuade J, Moran TH et al. Characterization of type A and type B CCK receptor binding sites in rat vagus nerve. Brain Res 1993; 623:161-166.
109. Ghilardi JR, Allen CJ, Vigna SR et al. Cholecystokinin and neuropeptide Y receptors on single rabbit vagal afferent ganglion neurons: site of prejunctional modulation of visceral sensory neurons. Brain Research 1994; 633:33-40.
110. Fraser KA, Davison JS. Cholecystokinin-induced c-fos expression in the rat brain stem is influenced by vagal nerve integrity. Exp Physiol 1992; 77:225-228.
111. Moran TH, Ameglio PJ, Schwartz GJ et al. Blockade of type A, not type B, CCK receptors attenuates satiety actions of exogenous and endogenous CCK. Am J Physiol 1992; 262:R46-R50.
112. Smith GP, Jerome C, Norgren R et al. Afferent axons in abdominal vagus mediate satiety effect of cholecystokinin in rats. Am J Physiol 1985; 249:R638-R641.
113. Crawley JN, Kiss JZ. Paraventricular nucleus lesions abolish the inhibition of feeding induced by systemical cholecystokinin. Peptides 1985; 6:927-935.
114. Crawley JN, Fiske SM, Durieux C et al. Centrally administered cholecystokinin suppresses feeding through a peripheral-type receptor mechanism. J Pharmacol Exper Ther 1991; 257:1076-1080.

115. Leslie RA, Gwyn DG, Hopkins DA. The central distribution of cervical vagus nerve and gastric afferent and efferent projections in the rat. Brain Res Bull 1982; 8:37-43.
116. Pagnani FD, Taveira da Silva AM, Hamosh P et al. Respiratory and cardiovascular effects of intraventricular cholecystokinin. Eur J Pharmacol 1982; 78:129-132.
117. Janssen PJJM, Gardiner SM, Compton AM et al. Mechanisms contributing to the differential haemodynamic effects of bombesin and cholecystokinin in conscious Long Evans rats. Brit J Pharmacol 1991; 102:123-134.
118. Hurle M, Morin-Surun MP, Foutz AS et al. Different targets involved in the effects of cholecystokinin on respiration. Eur J Pharmacol 1985; 118:87-96.
119. Szelenyi Z, Bartho L, Szekely M et al. Cholecystokinin octapeptide (CCK-8) injected into a cerebral ventricle induces a fever-like thermoregulatory response mediated by type B CCK-receptors in the rat. Brain Res 1994; 638:69-77.
120. Katsuura G, Hirota R, Itoh S. Cholecystokinin-induced hypothermia in the rat. Experientia 1981; 37:60.
121. South EH. Cholecystokinin reduces body temperature in vehicle-pretreated but not capsaicin-pretreated rats. Am J Physiol 1992; 263:R1215-R1221.
122. Wank SA, Harkins R, Jensen RT et al. Purification, molecular cloning, and functional expression of the cholecystokinin receptor from rat pancreas. Proc Nat Acad Sci U S A 1992; 89:3125-3129.
123. Debas HT, Farooq O, Grossman MI. Inhibition of gastric emptying is a physiological action of cholecystokinin. Gastroenterology 1975; 68:1211-1217.
124. Raybould HE, Tache Y. Cholecystokinin inhibits gastric motility and emptying via a capsaicin-sensitive vagal pathway in rats. Am J Physiol 1988: 255:G242-246.
125. Moran TH, McHugh PR. Gastric and nongastric mechanisms for the satiety effect of cholecystokinin. Am J Physiol 1988; 254:R628-R632.
126. Wettstein JG, Bueno L, Junien JL. CCK antagonists: pharmacology and therapeutic interest. Pharmac Ther 1994; 62:267-282.
127. Brooks AM, Grossman MI. Effect of secretin and cholecystokinin on pentagastrin-stimulated gastric acid secretion in man by peptide analogues of cholecystokinin. N Engl J Med 1970; 282:535-538.
128. Sandvik AK, Waldum HL. CCK-B (gastrin) receptor regulates gastric histamine release and acid secretion. Am J Physiol 1991; 260:G925-G928.
129. Soll AH, Amirian DA, Park J et al. Cholecystokinin potently releases somatostatin from canine fundic mucosal cells in short-term culture. Am J Physiol 1985; 248:G569-G573.
130. Lloyd KC, Raybould HE, Walsh JH. Cholecystokinin inhibits gastric acid secretion through type "A" cholecystokinin receptors and somatostatin in rats. Am J Physiol 1992; 263:G287-G292.

131. Konturek SJ, Kwiecien N, Obtulowicz W et al. Cholecystokinin in the inhibition of gastric secretion and gastric emptying in humans. Digestion 1990; 45:1-8.
132. Bradwejn J, Koszycki D, Payeur R. Role of cholecystokinin in irritable bowel syndrome. Am J Psychiatry 1991; 148:277.
133. Walker EA, Roy-Byrne PP, Katon WJ. Irritable bowel syndrome and psychiatric illness. Am J Psychiatry 1990; 147:565-572.
134. Uvnas-Moberg I, Arn I, Jonsson C-O et al. The relationships between personality traits and plasma gastrin, cholecystokinin, somatostatin, insulin, and oxytocin levels in healthy women. J Psychosom Res 1993; 37:581-588.
135. Brownson EA, Abbruscato TJ, Gillespie TJ et al. Effect of peptidases at the blood brain barrier on the permeability of enkephalin. J Pharm Exper Ther 1994; 270:675-680.
136. Blommaert AGS, Weng J-H, Dorville A et al. Cholecystokinin peptidomimetics as selective CCK-B antagonists: design, synthesis, and in vitro and in vivo biochemical properties. J Med Chem 1993; 36:2868-2877.
137. Rex A, Barth T, Voigt J-P et al. Effects of cholecystokinin tetrapeptide and sulfated cholecystokinin octapeptide in rat animal models of anxiety. Neurosci Lett 1994; 172:139-142.
138. Belzung C, Pineau N, Beuzen A et al. PD135158, a CCK-B antagonist, reduces "state," but not "trait" anxiety in mice. Pharmacol Biochem Behav 1994; 49:433-436.
139. Costall B, Domeney AM, Hughes J et al. Anxiolytic effects of CCK-B antagonists. Neuropeptides 1991; 19:65-73.
140. Derrien M, McCort-Tranchepain I, Ducos B et al. Heterogeneity of CCK-B receptors involved in animal models of anxiety. Pharmacol Biochem Behav 1994; 49:133-141.
141. Harro J, Vasar E. Evidence that CCK-B receptors mediate the regulation of exploratory behaviour in the rat. Eur J Pharmacol 1991; 193:379-381.
142. Harro J, Pold M, Vasar E. Anxiogenic-like action of caerulein, a CCK-8 receptor agonist, in the mouse: influence of acute and subchronic diazepam treatment. Naunyn-Schmiedeberg's Arch Pharmacol 1990; 341:62-67.
143. Hendrie CA, Neill JC, Shepherd JK et al. The effects of CCK-A and CCK-B antagonists on activity in the black/white exploration model of anxiety in mice. Physiol Behav 1993; 54:689-693.
144. Hsiao S, Katsuura G, Itoh S. Cholecystokinin tetrapeptide, proglumide, and open-field behavior in rats. Life Sci 1984; 34:2165-2168.
145. Hughes J, Boden P, Costall B et al. Development of a class of selective cholecystokinin type B receptor antagonists having potent anxiolytic activity. Proc Natl Acad Sci U S A 1990; 87:6728-6732.
146. Lemaire M, Bohme GA, Piot O et al. CCK-A and CCK-B selective receptor agonists and antagonists modulate olfactory recognition in male rats. Psychopharmacol 1994; 115:435-440.

147. Ervin FR, Palmour RM, Bradwejn J. A new primate model for panic disorder. Biol Psychiatry.1991; 29:430S.
148. Pavlasevic S, Bednar I, Qureshi GA et al. Brain cholecystokinin tetrapeptide levels are increased in a rat model of anxiety. NeuroReport 1993; 5:225-228.
149. Powell KR, Barrett JE. Evaluation of the effects of PD 134308 (CI-988), a CCK-B antagonist, on the punished responding of squirrel monkeys. Neuropeptides 1991; 19:75-78.
150. Rataud J, Darche F, Piot O et al. 'Anxiolytic' effect of CCK-antagonists on plus-maze behavior in mice. Brain Res 1991; 548:315-317.
151. Singh L, Field MJ, Hughes J et al. The behavioral properties of CI-988, a selective cholecystokinin-B receptor antagonist. Brit J Pharmacol 1991; 104:239-245.
152. Tessari M, Maraia G, Reggiani A. Possible role of CCK-B-receptors in anxiety: an "in vivo" study. Pharmacol Res 1992; 25:55-56.
153. Vasar E, Lang A, Harro J et al. Evidence for potentiation by CCK antagonists of the effect of cholecystokinin octapeptide in the elevated plus maze. Neuropharmacology 1994; 33:729-735.
154. Palmour RM, Ervin FR, Bradwejn J et al. Anxiogenic and cardiovascular effects of CCK-4 in monkeys are blocked by the CCK-B antagonist LY262691. Society for Neurosciences Abstract 1991; 637:1.
155. Crawley JN. Exploratory behavior models of anxiety in mice. Neurosci Biobehav Rev 1985; 9:37-44.
156. Crawley JN, Hays SE, Paul SM et al. Cholecystokinin reduces exploratory behavior in mice. Physiol Behav 1981; 27:407-411.
157. Jackson A, Tattersall D, Bentley G et al. An investigation into the discriminative stimulus and reinforcing properties of the CCK-B-receptor antagonist, L-365,260. Neuropeptides 1994; 26:343-353.
158. Rupniak NM, Schaffer L, Siegel P et al. Failure of intravenous pentagastrin challenge to induce panic-like effects in rhesus monkeys. Neuropeptides 1993; 25:115-119.
159. Harro J, Vasar E, Bradwejn J. CCK in animal and human research on anxiety. Trends Pharmacol Sci 1993; 14:244-249.
160. Harro J, Vasar E. Cholecystokinin-induced anxiety: how is it reflected in studies on exploratory behavior? Neurosci Biobehav Rev 1991; 15:473-477.
161. Hughes J, Hunter JC, Woodruff GN. Neurochemical actions of CCK underlying the therapeutic potential of CCK-B antagonists. Neuropeptides 1991; 19:85-89.
162. Lydiard RB. Neuropeptides and anxiety: focus on cholecystokinin. Clin Chem 1994; 40:315-318.
163. Ravard S, Dourish CT. Cholecystokinin and anxiety. Trends Pharmacol Sci 1990; 11:271-273.

CHAPTER 7

# CCK, Addiction and Anxiety

Franco Vaccarino

## DRUG DEPENDENCE AND WITHDRAWAL: ROLE OF CCK ANXIOGENIC PROCESSES

The term "addiction" is generally taken to refer to dependent patterns of drug self-administration. Two commonly accepted categories of drug dependence are physical and psychological dependence. Physical dependence refers to a state in which the body has adjusted to the presence of the drug such that when the drug is withdrawn, there are physical symptoms of withdrawal often involving discomfort and pain.[1] Psychological dependence, in contrast, is characterized by preoccupation with the drug's emotional and mental effects, and by persistent craving for these effects.[1] These emotional and psychological effects are typically associated with states of positive affect and reflect the rewarding and incentive value of the drug. It is worth noting that psychological dependence may also involve withdrawal reactions to the extent that the drug's psychological effects induce opposing psychological processes. For example, while a drug's actions may be associated with euphoric psychological effects, termination of the drug's actions may be associated with dysphoric effects or depressed mood. Thus, in addition to referring to the incentive value of the drug in question, psychological dependence may also involve negative mood states associated with drug withdrawal.[1]

*Cholecystokinin and Anxiety: from Neuron to Behavior,* edited by Jacques Bradwejn and Eero Vasar. 

Anxiety is a central feature of withdrawal processes. For this reason, one focus of this chapter will be to review evidence that CCK may be involved in mechanisms underlying anxiety processes associated with withdrawal states. There is also considerable evidence to suggest that CCK interacts with mechanisms known to mediate the incentive or rewarding value of drugs of abuse. This raises the additional possibility that CCK plays a role in mechanisms underlying the rewarding properties of certain dependence producing drugs.

## DRUG WITHDRAWAL

Although there is a long list of drugs known to produce withdrawal reactions following long term use, this chapter will only highlight those that may be relevant to anxiety and CCK.

### ALCOHOL

Cessation of alcohol intake following long term use produces a variety of withdrawal symptoms.[1] These symptoms appear in three consecutive phases. Phase 1 symptoms (can begin within a few hours) include: severe tremulousness (the shakes), profuse perspiration, weakness, agitation, nausea, vomiting, anorexia, abdominal cramps, hyperreflexia and hallucinations. Phase 2 symptoms (begin within 24 hours) are characterized by grand mal seizures. Phase 3 is the most protracted phase and is characterized by delirium tremens (DTs). During this phase the person is severely agitated and often appears confused and disoriented. The individual can also experience terrorizing auditory and tactile hallucinations which can produce delusions and violent behavior.

### PSYCHOSTIMULANTS

Long term use of psychostimulants, such as cocaine or amphetamine, is associated with a variety of withdrawal symptoms including: anxiety, agitation, mood swings, toxic psychosis, hypersensitivity to sensory stimuli (sometimes producing hallucinations), sleep and eating disorders and depression. Withdrawal from long term use of nicotine, classified as a mild CNS stimulant, is associated with depression, anxiety, restlessness, irritability, sleep disturbances, constipation, sweating and long lasting craving.[1]

### OPIATES

Powerful physical dependence occurs quickly following regular high dose use of opiates such as heroin or morphine. Early symptoms of withdrawal sickness include: nasal discharge, uncontrollable yawning, watering eyes, agitated sleep, loss of appetite, depression and tremor. At peak intensity, symptoms include: alternating chills and flushing, insomia, violent yawning and sneezing, vomiting nausea, aching bones, muscles and joints and uncontrollable kicking movements. Psychological withdrawal symptoms include: chronic depression, anxiety, insomnia, agitation and long periods of intense craving for the drug.[1,2]

### BENZODIAZEPINES

The risk of physical dependence arising from long term use of benzodiazepines is higher than originally thought. The most common withdrawal symptoms are rebound anxiety involving bodily symptoms of anxiety and panic, insomnia and nightmares. Other symptoms include restlessness, shakiness, tremors and hypersensitivity to light, sound, touch and odors. Other symptoms include: burning, itching and prickling sensations on the skin, mood changes, apathy, irritability, impaired memory and concentration, flu-like symptoms and loss of appetite.[1,3,4]

## DRUG WITHDRAWAL AND ANXIETY

Although the specific profile of withdrawal symptoms differs across the different drug classes, there are certain common features. Most importantly for this chapter is the fact that anxiety is a prominent feature in withdrawal reactions. Indeed many of the psychological dependence signs of withdrawal states may also represent anxiogenic features of behavior (e.g., agitation, depression, insomnia).

Numerous animal[5] and human[1,2,6] studies support the notion that anxiety is a common feature of drug dependence and withdrawal states. Brown et al[7], for example, examined anxiety characteristics in 171 male alcoholics during inpatient treatment as well as for 3 months following treatment. Their results indicated that state anxiety scores were associated with abstinence and were highest during the first week of treatment. During the second and third weeks, anxiety symptoms tended to normalize, although symptoms continued to decrease with each week of continued abstinence. It is important to note that withdrawal-induced anxiety is not drug specific and is evidenced in other drug groups including psychostimulants and opiates. For example, Swift and Stout[8] showed that anxiety scores correlated with subjective withdrawal symptoms in 32 opiate-dependent patients. Thus, withdrawal-induced anxiety decreases as the withdrawal proceeds and withdrawal symptoms decline. These results are consistent with other reports and indicate that anxiety symptoms are a central feature of drug withdrawal in chronic drug abusers.

Animal studies are consistent with the human findings and indicate that withdrawal from long term exposure to drugs such as cocaine, nicotine, alcohol and opiates is associated with increases in anxiogenic behaviors. This has been demonstrated using a variety of tests for anxiogenic behavior including: the elevated plus maze, social interaction, exploratory behavior, fear potentiated startle and punished responding paradigms.[9-14]

The central role played by anxiety processes during withdrawal states is underlined by the fact that benzodiazepine anxiolytics are effective therapeutic agents in controlling drug withdrawal states in both animals and humans.[15,16] It is worth noting, however, that benzodiazepines are themselves associated with withdrawl symptoms following long term exposure.[17] Importantly, the behavioral characteristics and pharmacological profile of diazepam and alcohol withdrawal are similar,[1,18] suggesting an overlap between the neurochemical mechanisms mediating the anxiogenic effects of alcohol and benzodiazepine withdrawal.

## CCK, DRUG WITHDRAWAL AND ANXIETY

The above discussion indicates that anxiety mechanisms are a central feature of drug dependence and more specifically, withdrawal processes. Before describing evidence implicating CCK in dependence and withdrawal-induced anxiety, it is important to note that there is considerable evidence suggesting a strong link between CCK and anxiety mechanisms. This evidence comes from a variety of sources. First, in both man and animals, treatment with CCK or CCK agonists induce anxiogenic responses. Second, these anxiogenic responses appear to be largely related to CCK actions at the CCK-B receptor since CCK-B agonists are the most effective anxiogenics, and since CCK-B, but not CCK-A, antagonists, reverse the anxiogenic effects of CCK. Third, animal studies have shown that anxiogenic states are associated with elevated levels of CCK and CCK receptors in a variety of brain regions. Finally, CCK-B antagonists alone have been found to have direct anxiolytic effects on basal anxiogenic behavior.[19,20]

Of particular relevance here are findings specifically implicating CCK in anxiogenic processes associated with drug dependence and withdrawal. Perhaps the most direct evidence supporting a role for CCK in anxiogenic behaviors associated with withdrawal comes from Costall et al.[21] They tested mice with chronic administration of diazepam (10 mg/kg/day for 7 days), nicotine (0.1 mg/kg/day for 7 days), cocaine (1.0 mg/kg/day for 14 days) and alcohol (8% w/v in drinking water for 14 days). These drug administration schedules produce dependence as evidenced by the presence of withdrawal symptoms following drug termination. Costall et al[21] found that termination of each of these chronic drug treatments was associated with anxiogenic behavior in the white/black exploratory test, a reliable model of anxiogenic behavior. Moreover, they found that pretreatment with the selective CCK-B antagonists, PD135158 and PD134308, attenuated the expression of withdrawal-induced anxiogenesis. This effect was found following all dependence-producing drug classes tested, suggesting that CCK-anxiogenic mechanisms represent a common substrate for withdrawal processes associated with different drug classes. Costall et al[21] further reported that chronic treatment with CCK-B antagonists was not associated with any signs of dependence. Thus, the anxiolytic profile of CCK-B antagonists may not share the dependence producing properties of benzodiazepine anxiolytics.

Further support for the role of CCK mechanisms in withdrawal processes comes from Rasmussen et al[14] who showed that, in rats, a different CCK-B antagonist (LY288513) attenuated the potentiated startle response found following withdrawal from chronic diazepam treatment. These results provide converging evidence supporting the results of Costall et al[21] by showing withdrawal-specific anxiolytic effects of CCK-B antagonists in a different species, a different model of anxiogenic behavior and a different CCK-B antagonist.

A number of studies examining levels of CCK and expression of CCK mRNA in the brain following acute and chronic administration of morphine, have provided information which further implicate CCK in mechanisms underlying opiate dependence and withdrawal. It has been found that acute and chronic morphine treatment results in increased CCK and CCK mRNA content in a variety of brain regions,[22,23] and that CCK antagonists or antiserum prevent the development of tolerance to chronic morphine treatment.[24,25] Of particular interest here are findings showing that increases in CCK mRNA were highest following chronic exposure to morphine.[22] These results are generally consistent with the notion that increases in CCK activity are associated with the development of dependence and withdrawal-related anxiogenesis. In addition, these results provide indirect evidence implicating CCK in mechanisms subserving opiate dependence and withdrawal.

## CCK-DOPAMINE INTERACTIONS: RELEVANCE TO REWARD PROCESSES

The above discussion has focused on withdrawal processes associated with chronic exposure to addictive drugs. However, prior to long term exposure and the full development of drug dependence, the initial phases of drug abuse begin with the rewarding properties of drugs of abuse. It is the rewarding properties of addictive drugs which promote the initial drug seeking behavior (i.e., reinforce drug taking behavior) and which give addictive drugs their abuse potential.

Psychologically, the rewarding properties of drugs are reflected in their ability to induce states of positive affect and general well being. In addition to being reflected in direct self-administration behavior (i.e., drug taking behavior), the rewarding properties of drugs of abuse are reflected in their behavioral activating properties. In both man and other animals, the rewarding properties of addictive drugs are often paralleled by increased feelings of energy and behavioral activation. The behavioral activation induced by addictive drugs is thought to be a behavioral manifestation of an increased reward state.[26,27]

Studies aimed at uncovering the neurochemical mechanisms underlying the rewarding properties of abused drugs have repeatedly demonstrated the importance of dopamine (DA) neurotransmission (particularly mesolimbic DA). This section will provide a brief overview of the central role played by DA in reward processes and further examine the contribution of CCK to reward-relevant behaviors. CCK is of special interest for three reasons. First, it has neurotransmitter-like properties. Second, it coexists with dopamine in a significant subpopulation of mesolimbic DA neurons and is released in the nucleus accumbens (N.Acc) where it influences DA function. Third, CCK in the N.Acc also derives from non-DA neuronal sources. This information is reviewed by Crawley[28] and Vaccarino.[29] Mechanisms to account for DA-CCK interactions underlying drug reward and related processes will be proposed.

## DRUG REWARD AND DOPAMINE

Much of the literature implicating DA in drug reward mechanisms comes from studies examining the neurochemical mechanisms underlying the rewarding properties of psychostimulant drugs. The psychostimulants, cocaine and amphetamine (AMPH), are well known for their abuse potential. In this regard, the critical mechanism of action for cocaine is its ability to block DA reuptake, while the primary mode of action for AMPH is the stimulation of DA release.[30] These mechanisms allow for increased synaptic transmission of DA.

Over the years considerable evidence has emerged indicating that the mesolimbic dopamine (DA) system is a critical substrate for the rewarding and behavioral activating properties of the psychostimulant drugs, cocaine and AMPH.[26,27,31-35] Essentially, these studies show that the rewarding properties of a variety of drugs of abuse are linked to activation of DA systems. Moreover, the studies show that the rewarding properties of psychostimulant drugs are dependent on the integrity of the mesolimbic DA system, such that when this system is compromised, drug reward is attenuated and when this system is activated, drug reward is enhanced.

Although psychostimulants have been emphasized in studies of DA involvement in drug reward, it is important to note that DA function has also been linked to the rewarding properties of non-DAergic drugs of abuse, such as opiates, nicotine and alcohol.[26] In these cases, indirect activation of mesolimbic DA activity is believed to, at least in part, contribute to the rewarding properties of these drugs.[26,35-40]

## FOOD REWARD AND DOPAMINE

In addition to being a critical substrate for psychostimulant reward, the mesolimbic DA system may also be the substrate through which natural reinforcers, such as food, derive their rewarding properties. Evidence has accumulated that supports the involvement of central DAergic reward mechanisms in feeding. Radhakishun et al,[41] utilizing on-line in vivo microdialysis, observed that the initiation of eating increased the release of DA in the N.Acc and the level of N.Acc DA remained elevated during the entire consummatory episode. The termination of feeding was accompanied by a gradual decrease in the release of DA and an eventual return to the pre-prandial basal level of release.

Hernandez and Hoebel[42,43] have further demonstrated, using in vivo microdialysis and post-mortem analysis techniques, that operant responding for food reward, lateral hypothalamic stimulation and psychostimulant reward all increased extracellular DA concentrations in the N.Acc. These findings led the authors to conclude that psychomotor stimulants increase DA activity in a neural system that both mediates reinforcement and is activated by eating.

Further evidence for the involvement of DA reward mechanisms in feeding comes from studies showing that, at low doses, AMPH facilitates feeding.[44-49] Moreover, the N.Acc is an important substrate for this effect.[46,50,51] AMPH-stimulated food intake is believed to be mediated by the increased rewarding properties of food induced by elevated DA activity.[46] Interestingly, the increased food intake induced by intra-N.Acc AMPH has been demonstrated to be sucrose-specific in rats presented with a choice of various chow/sucrose combinations. This indicates that, in the context of feeding, the AMPH reward signal appears to be preferentially coupled with sucrose.[26]

## INDIVIDUAL DIFFERENCES IN RESPONSIVITY TO PSYCHOSTIMULANT DRUGS

Recent evidence has extended the role of mesolimbic DA to account for individual differences in behavioral responsivity to psychostimulant drugs.

### Psychostimulants and Locomotor Activity

Piazza et al[52] found that rats show individual differences in their locomotor response to a novel environment. Based on these differences, they separated rats into low responders (LR) and high responders (HR), and found that LR exhibited less locomotor activation in response to AMPH administration than HR. These findings have been replicated by Hooks et al.[53]

Both Piazza et al[52] and Hooks et al[53] have suggested that intrinsic differences in the mesolimbic DA system may underlie the observed individual differences in locomotor responsiveness to both a novel environment and psychostimulant administration. This notion is supported by the observation that HR exhibit a higher DOPAC/DA ratio in the N.Acc than LR in response to a novel environment and under basal conditions.[54] Recent in vivo microdialysis results are consistent with this concept. Hooks et al[53,55] have shown that HR release more DA in the N.Acc than LR following treatment with cocaine, as well as under basal conditions.

### Sensitization to Amphetamine-Induced Locomotion

Repeated daily administrations of AMPH or cocaine produce profound behavioral sensitization, as evidenced by increased levels of locomotor activity and more intense stereotypy.[56-59] As in the case of acute administrations, there are individual differences in the response to repeated administrations of AMPH. Hooks et al[60] found that HR exhibited sensitization of their locomotor response to repeated AMPH administrations, while LR did not. These results are consistent with recent reports from our laboratory.[61]

### Amphetamine Self-Administration

Piazza et al[52] reported on individual differences in the acquisition of AMPH self-administration. They examined the acquisition of (intravenous) AMPH self-administration in two groups of rats as a function of their locomotor response to a novel environment. They found that HR readily self-administered AMPH while LR did not. Additionally, they found that repeated injections of AMPH, induced the acquisition of self-administration in LR. Piazza et al[52] concluded that the observed individual differences in the vulnerability to self-administer AMPH may reflect intrinsic differences in the neural mechanisms mediating responsiveness to AMPH. This notion is supported by microdialysis results showing the LR have lower N.Acc DA levels than HR.[53,54,62] It is of interest that individual differences in predisposition to self-administer morphine have also been reported.[63,64]

### Amphetamine-Induced Feeding

Recent evidence indicates that there are reliable individual differences in the baseline feeding behavior of rats.[65-67] These individual differences can predict the direction of feeding response to low doses of AMPH. More specifically, it has been reported that AMPH increased sugar intake in rats with low baseline intake and inhibited sugar intake in animals with high baseline intake.[66,67]

Similar to Piazza et al[52] and Hooks et al,[53] it was proposed[65,66] that high baseline feeders have higher N.Acc DA activity, and hence higher sugar intake, than low baseline feeders. Thus, administration of a low dose of AMPH to high baseline feeders would be expected to induce a state of DA hyperstimulation resulting in anorexia.[51,68] A low dose of AMPH administered to low baseline feeders, on the other hand, would be expected to produce a level of DAergic activation sufficient to stimulate feeding.[51]

Recent evidence[67] supports the notion that low and high feeders differ in their level of endogenous N.Acc DA activity. Intra-N.Acc microinjections of alpha-flupenthixol (a non-selective DA antagonist) attenuates the naturally elevated sugar intake of high feeders, as well as the elevated sugar intake in low feeders treated with AMPH (low dose). Thus, the high feeders resemble the low feeders treated with AMPH.

## RELATION BETWEEN REWARD, LOCOMOTION AND FEEDING

Taken together, the above overview indicates that increased DA transmission in the mesolimbic DA system is a critical mechanism for the rewarding and locomotor activating properties of the psychostimulant drugs, cocaine and AMPH, as well as natural reinforcers, such as food. It should be noted that in addition to feeding, the mesolimbic DA system has been implicated in sexual reward,[69,70,71] as well as other natural reinforcers.[27] The association of these different behavioral properties with N.Acc DA, suggests a functional commonality which has theoretical and applied significance.[26,72] Extending earlier notions of Glickman and Schiff,[27] it has been suggested that DAergic reward signals represent the activation of species typical behavioral systems.[27] Vaccarino, Schiff and Glickman[27] viewed locomotor activation as a common denominator of naturally motivated appetitive behaviors, reflecting the approach behavior common to all these behaviors.[27] Wise and Bozarth[26] have further discussed this notion with respect to its application in the understanding of drug addiction processes.[73]

In addition to accounting for the rewarding properties of other behaviors (i.e., locomotor activity and feeding), as has already been mentioned, the involvement of DA in reward processes has also been extended to account, at least in part, for the rewarding properties of non-DAergic drugs such as alcohol, nicotine and opiates. Like psychostimulants, these non-DAergic addictive drugs also have stimulant properties (particularly at low doses) and these properties are believed to underlie their motivational effects on drug seeking behavior.[26] The manifestation of stimulant properties of these drugs is believed to represent the contribution of DAergic mechanisms.[26]

## CCK-DOPAMINE INTERACTIONS IN REWARD-RELEVANT BEHAVIORS

Immunohistochemical studies have shown that in addition to containing DA, a subset of mesolimbic DA neurons (more than 40%) also contain CCK.[74-79] While a number of different forms of CCK have been localized in the brain, it appears that the sulfated C-terminal CCK octapeptide is the predominant active form in the brain.[80] Unless otherwise stated, the term "CCK" will refer to the sulfated octapeptide form of CCK. CCK is synthesized de novo,[81] released from synaptic terminals in a calcium dependent manner[82] and demonstrates specific binding characteristics[83] indicating a neurotransmitter-like profile. Studies examining the electrophysiological, biochemical and behavioral properties of CCK support the notion that CCK acts as a neuromodulator/ neurotransmitter in the brain. While CCK is found in numerous brain regions to varying degrees, the present discussion will examine the significance of CCK as it relates to the mesolimbic DA system, with particular emphasis on the N.Acc terminal region.

The co-existence of CCK and DA in the mesolimbic pathway has given rise to a number of studies aimed at investigating the role of CCK in mesolimbic DA mediated behaviors. The following section provides an overview of findings which highlight the neurobiological and behavioral complexities underlying CCK-DA interactions.

## PSYCHOSTIMULANTS AND LOCOMOTOR ACTIVATION

The first behavioral studies to examine the functional significance of N.Acc CCK focused on DA-mediated locomotor activity. Initial studies yielded mixed results. Crawley et al[84,85] reported that intra-N.Acc CCK potentiated the locomotor activating effects of intra-N.Acc DA or systemic apomorphine treatments. In contrast, Van Ree et al[86] and Weiss et al[87] reported that intra-N.Acc CCK antagonized the locomotor potentiating effects of intra-N.Acc DA and systemic AMPH, respectively. Thus, N.Acc CCK has been reported to potentiate and attenuate DA-mediated locomotor activation.

The opposing effects of N.Acc CCK on DA-mediated behavior appear to depend, at least in part, on the rostral-caudal location of the intra-N.Acc CCK treatment. Vaccarino and Rankin[88] reported that CCK potentiated AMPH-induced locomotor activity when CCK was injected into the caudal N.Acc, and blocked AMPH-induced locomotor activity when CCK was injected into the rostral N.Acc. Consistent with this finding, re-examination of earlier findings[84-86] indicated that the DA agonist-like effects of CCK are associated with CCK actions in the caudal N.Acc, while the DA antagonist-like effects of CCK are associated with CCK actions in the rostral N.Acc.

Since the post-synaptic effects of CCK in the N.Acc are generally excitatory and opposite those of DA, it is reasonable to hypothesize that the DA agonist-like behavioral effects of CCK in the caudal N.Acc are mediated by the pre-synaptic DA stimulating properties of CCK in this region. Based on the Blaha et al[89] electrochemical detection findings, optimal conditions for CCK-induced potentiation of DA mediated behaviors would require low levels of exogenous CCK concentrations. In addition, the most likely time course for the excitatory effects of CCK on DA-mediated behaviors would be immediately post-injection. Indeed, the results of Crawley et al[84,85] support this view. They report that the DA potentiating effects of intra-N.Acc CCK are expressed during the first 10 minutes following intra-N.Acc injections of low doses of CCK. Recent unpublished results from our laboratory are also consistent with these findings and show that the percentage of locomotor activity occurring immediately after CCK injections into the caudal N.Acc decreases as CCK dose increases.

Investigations of the effects of CCK antagonism are consistent with the hypothesized DA agonist-like and antagonist-like roles of CCK in the caudal and rostral N.Acc, respectively. Crawley[28,90] has reported that, in the rostral N.Acc, CCK-B, but not CCK-A, antagonists block CCK-induced attenuation of DA-mediated locomotion. In contrast, in the caudal N.Acc, CCK-A, but not CCK-B, antagonists block CCK-induced potentiation of DA-mediated locomotion. In consideration of the neurochemical and pharmacological properties of CCK, it is possible that the CCK-A mediated, stimulatory effects of CCK on DA release in the caudal N.Acc,[91,92] contribute to the DA agonist-like effects on locomotor activity, while the CCK-B mediated, inhibitory effects of CCK on DA release in the rostral N.Acc,[92] contribute to the DA antagonist-like effects of CCK on locomotor activity.

## BRAIN STIMULATION AND DRUG REWARD

It has been reported that intra-N.Acc CCK injections exert strong modulatory effects on various reward-relevant behaviors including intravenous cocaine self-administration and brain stimulation reward.[93,94,95,96] With regard to intravenous cocaine self-administration, microinjections of low doses of CCK into the N.Acc caused a dose-dependent increase in cocaine self-administration. The dose-response and time-course profile was consistent with a neuroleptic-like reward-blocking effect.[97] Further support for the notion that these CCK effects reflected reward blockade, was the finding that intra-N.Acc CCK microinjections attenuate brain stimulation reward.[95]

The extent to which the site of injection within the N.Acc was a factor in the cocaine self-administration study or in the brain stimulation reward study was not systematically investigated. Based on the locomotor activity findings,[88] it is likely that these DA antagonist-like effects reflect CCK actions in the rostral N.Acc. This is consistent with the rostral-caudal differences in CCK effects on brain stimulation reward reported by De Witte et al.[98]

In the only study to date to investigate the effects of N.Acc CCK antagonism on reward functions, site-dependency was also found to be an important consideration. Injections of the nonselective CCK receptor antagonist, proglumide, into the caudal and rostral N.Acc produced an attenuation and potentiation, respectively, of brain stimulation reward derived from the ventral tegmental area.[99] This finding is consistent with the locomotor activity findings and further reinforces the notion that reward and locomotor activation rely on a common substrate.[27]

Recently using the conditioned place preference[100] and conditioned reward[101] paradigms, further evidence indicates that endogenous

CCK actions on CCK-B receptors play a role in morphine and AMPH reward.[100,102] More specifically, the CCK-B antagonist, L-365,260, was found to potentiate morphine and AMPH-induced conditioned place preference, as well as AMPH potentiation of conditioned reward. These results suggest that endogenous CCK exerts inhibitory effects on mechanisms underlying AMPH-induced place preference and conditioned reward. This is consistent with the reward-blocking actions of N.Acc CCK, further suggesting that the rostral N.Acc is a good candidate for the site of action of this effect.

## INDIVIDUAL DIFFERENCES IN RESPONSIVENESS TO AMPHETAMINE

### Locomotor Activity

It has recently been found that a CCK-B, but not CCK-A, receptor antagonist, dose-dependently increased AMPH-induced locomotor activity in animals with low, but not high, baseline activity levels.[103] These results suggest that individual differences in baseline locomotor activity levels may be an important variable for determining the behavioral effectiveness of CCK and CCK antagonists. Moreover, this particular result raises the possibility that elevated endogenous CCK-B receptor activation contributes to the lowered activity level in low responders.

### Food Reward

Extending the importance of individual differences to the feeding paradigm, Sills and Vaccarino[65-67] found that microinjections of CCK into the rostral N.Acc, blocked sugar intake in high, but not low, sugar eaters. Additionally, intra-N.Acc CCK administration blocked the elevated sugar intake in low baseline feeders treated with AMPH. These results are very similar to recent findings from our laboratory examining the effects of intra-N.Acc neuroleptic treatment on sugar intake.[67]

More recently it was found that the CCK-B antagonist, L-365,260, exerts DA agonist-like effects on both low and high feeders.[102] This result complements the latter findings and further suggests that endogenous CCK exerts tonic neuroleptic-like effects on N.Acc DAergic mechanisms contributing to elevated food intake and reward.

## PROPOSED MECHANISMS FOR CCK-DOPAMINE INTERACTIONS IN DRUG REWARD

Based on the preceding discussion, the following mechanisms are proposed to account for CCK-DA interactions underlying reward-related functions:

a. Exogenous CCK actions at the CCK-A receptor facilitate the actions of DA and DAergic reward processes. These effects likely reflect increased DA release induced via activation of CCK-A receptors located on DA terminals in the N.Acc, particularly in the caudal N.Acc.
b. Exogenous CCK actions at the CCK-B receptor antagonize the actions of DA and DAergic reward processes. These effects likely reflect the actions on post-synaptic CCK receptors in the N.Acc and inhibition of DA release via activation of CCK-B receptors located on DA terminals in the rostral N.Acc.
c. Endogenous CCK activity in the N.Acc contributes to mechanisms mediating the behavioral effects of psychostimulant drugs, and possibly other drugs of abuse. More specifically, endogenous activation of CCK-B receptors exerts an inhibitory influence on psychostimulant reward and locomotor activation.
d. The effects of CCK and CCK antagonists on psychostimulant-induced behaviors are influenced by individual differences in baseline behavioral state. This further suggests that endogenous CCK activity may contribute to the neurobiological substrates underlying individual differences in behavioral responsivity to psychostimulant drugs.
e. It is important to note that the present purported mechanisms of action for CCK seem to generalize across three behavioral paradigms: iv psychostimulant reward, psychostimulant-induced locomotor activation and psychostimulant-induced feeding. This suggests that CCK is acting at a motivational level which cuts across behavioral dimensions. The generalizability of CCK and DA actions across these three different behavioral dimensions is consistent with the notion that these neurochemical signals are influencing processing of reward-related information.

As is evidenced in the preceding review, the behavioral significance of CCK-DA interactions are complex and depend on a number of factors including: CCK receptor subtype, pre- vs post-synaptic site of action, rostral-caudal location within the N.Acc and individual differences in the baseline behavioral state of the animal. These complexities have given rise to diverse, and sometimes inconsistent, findings concerning the functional significance of CCK in the expression of DA-mediated behaviors. It is hoped that the present synthesis of information will provide a useful working framework for examination of the role of CCK in DA reward and related processes.

## POSSIBLE RELEVANCE OF CCK REWARD EFFECTS TO ANXIETY MECHANISMS

The involvement of CCK in both anxiety and reward processes raises the intriguing possibility that a component of the anxiogenic properties of CCK reflects the contribution of CCK actions in the N.Acc and effects on reward processes. In light of the anti-DAergic effects of CCK acting through the CCK-B receptor, it is interesting to speculate that an element of anxiogenic profile of CCK involves the ability of CCK to depress DAergic activity. An extension of this notion would suggest that an element of the anxiolytic profile of CCK-B antagonists involves the DA potentiating effects of CCK-B antagonists.

This conceptualization has a number of implications concerning dependence, reward and anxiety. For example, there is evidence to suggest that termination of psychostimulant drug actions, even after short term use is associated with a rebound decrease in DAergic activity.[104] It may be, then, that some of the psychological dependence signs (e.g., craving, depression, anxiety) are mediated by this rebound decrease in DAergic activity. In fact, this notion has been proposed by Koob[105] in his opponent process model of drug dependence. The present discussion of CCK involvement in reward mechanisms would further suggest that elevated CCK activity acting at the CCK-B receptor could contribute to mechanisms underlying decreased DA activity. Moreover this elevated CCK activity could contribute to this "DA withdrawal" state, and possibly the anxiogenic features of this state.

In light of the opposing actions of CCK acting at CCK-A receptors, CCK actions at the CCK-A receptor may play a very different role which may or may not be associated with anxiety. For example, it may be that CCK actions at the CCK-A receptor contribute to the acquisition of psychostimulant reward by enhancing DA function under conditions of drug-induced DA activation. It would be particularly interesting to examine the extent to which CCK-A receptors play a role in the sensitization to psychostimulant drugs, a process thought to be critical in the development of drug reward.

## REFERENCES


1. Jacobs MR, Fehr KO, eds. Drug and Drug Abuse-A Reference Text. 2nd ed. Toronto: Alcoholism and Drug Addiction Research Foundation, 1987.
2. Martin WR, Fraser HF. A comparative study of physiological and subjective effects of heroin and morphine administered intravenously in post-addicts. J Pharmacol Exp Ther 1961; 133:388-399.
3. Ladewig D. Dependence Liability of Benzodiazepines. Drug Alcohol Depend 1984; 13:139-149.
4. Petursson H, Lader M. Dependence on Tranquilizers. Oxford University Press, New York 1984.
5. Goeders NE. Potential involvement of anxiety in the neurobiology of cocaine. In: Kalivas PW, Samson HH, eds. The Neurobiology of Drug and Alcohol Addiction Ann NY Acad Sci; 1992:654.
6. Gawin FH, Ellinwood EH. Cocaine and other stimulants: actions abuse and treatment. New Eng J Med 1988; 318:1173-1182.
7. Brown SA, Irwin M, Schuckit MA. Changes in anxiety among abstinent male alcoholics. J Stud Alcohol 1991; 52:55-61.
8. Swift RM, Stout RL. The relationship between craving, anxiety, and other symptoms in opioid withdrawal. J Subst Abuse 1992; 4:19-26.
9. File SE, Baldwin HA, Aranko K. Anxiogenic effects in benzodiazepine withdrawal are linked to the development of tolerance. Brain Res Bull 1987; 19:607-610.
10. McMillan DE, Leander JD. Chronic chlordiazepoxide interactions onpunished and unpunished behavior. J Pharmacol Exp Ther 1978; 207:515-520.
11. File SE, Baldwin HA, Hitchcott PK. Flumazenil but not nitrendipine reverses the increased anxiety during ethanol withdrawal in the rat. Psychopharmacol 1989; 98:262-264.
12. Costall B, Jones BJ, Kelly ME et al. Exploration of mice in black and white test box: validation as a model of anxiety. Pharmacol Biochem Behav 1989; 32:777-785.
13. File SE. Animal tests of anxiety. Recent Adv Neuropsychopharmacol 1981; 31:241-251.

14. Rasmussen K, Helton DR, Berger JE et al. The CCK-B antagonist LY288513 blocks effects of diazepam withdrawal on auditory startle. NeuroReport 1993; 5:154-156.
15. Crowley TJ. Clinical issues in cocaine abuse. In: eds Fisher S, Raskin A, Uhlenhuth EH. Cocaine: Clinical and Biobehavioral Aspects. Oxford University Press, New York 1987:193-211.
16. Jaffe JJ. Drug addiction and drug abuse. In: eds Gilman AG, Rall TW, Nies AS, Taylor P. The Pharmacological Basis of Therapeutics. 8th ed. Toronto: Pergamon Press, 1990:522-573.
17. Little HJ. The benzodiazepines: anxiolytic and withdrawal effects. Neuropeptides 1991; 19:11-14.
18. File SE, Zharkovsky A, Hitchcott PK. Effects of nitrendipine, chlordiazepoxide, flumazenil and baclofen on the increased anxiety resulting from alcohol withdrawal. Prog Neuropsychopharmacol Biol Psychiatry 1992; 16:87-93.
19. Harro J, Vasar E, Bradwejn J. Cholecystokinin in animal and human research on anxiety. Trends Pharmacol Sci 1993; 14:244-249.
20. Lydiard RB. Neuropeptides and anxiety: focus on cholecystokinin. Clin Chem 1994; 40:315-318.
21. Costall B, Domeney AM, Hughes J et al. Anxiolytic effects of CCK-B antagonists. Neuropeptides 1991; 19:65-73.
22. Ding XD, Bayer BM. Increases of CCK mRNA and peptide in different brain areas following acute and chronic administration of morphine. Brain Res 1993; 625:139-144.
23. Faris PL, Beinfeld MC, Scallet AC et al. Increase in hypothalamic cholecystokinin following acute and chronic morphine. Brain Res 1986; 367:405-407.
24. Ding XZ, Fan SG, Han JS. Reversal of tolerance to morphine but no potentiation of morphine-induced analgesia by antiserum against cholecystokinin octapeptide. Neuropsychopharmacol 1986; 25:1155-1160.
25. Watkins LR, Kinscheck IB, Mayer DJ. Potentiation of opiate analgesia and apparent reversal of morphine tolerance by proglumide. Science 1984; 224:395-396.
26. Wise RA, Bozarth MA. A psychomotor stimulant theory of addiction. Psychol Rev 1987; 94:469-492.
27. Vaccarino FJ, Schiff BB, Glickman SE. A biological view of reinforcement. In: eds Klein SB, Mowrer RR. Contemporary Learning Theories Hillsdale, NJ, Erlbaum 1989: 111-142.
28. Crawley JN. Cholecystokinin-dopamine interactions. Trends Pharmacol Sci 1991; 121:232-236.
29. Vaccarino FJ. Nucleus accumbens dopamine-CCK interactions in psychostimulant reward and related behaviors. Neurosci Biobehav Rev 1994; 18:207-214.
30. Kuczenski R. Biochemical actions of amphetamine and other stimulants. In: ed Creese I. Stimulants: Neurochemical, Behavioral and Clinical Perspectives, New York: Raven, 1983:31-61.

31. DiChiara G, Imperato A. Drug abused by humans preferentially increase synaptic dopamine concentrations in the mesolimbic system of freely moving rats. Proc Natl Acad Sci USA 1988; 85:5274-5278.
32. Kalivas PW, Stewart J. Dopamine transmission in the initiation and expression of drug and stress-induced sensitization of motor activity. Brain Res. Rev 1991; 16:223-244.
33. Koob GF, Vaccarino FJ, Amalric A et al. Positive reinforcement properties of drugs: Search for neural substrates. In: eds Engel J et al. Brain Reward Systems and Abuse. New York: Raven Press, 1987:35-50.
34. Robbins TW, Everitt BJ. Functional Studies of Central Catecholamines. Int. Rev Neurobiol 1982; 23:303-365.
35. Wise RA, Rompre P. Brain dopamine and reward. Ann Rev Psychol 1989; 40:191-225.
36. Di Chiara G, Imperato A. Drugs abused by humans preferentially increase synaptic dopamine concentrations in the mesolimbic system of freely moving rats. Proc Natl Acad Sci 1988; 85:5274-5278.
37. Di Chiara G, Imperato A. Opposite effects of μ and κ-opiate agonists on dopamine-release in the nucleus accumbens and in the dorsal caudate of freely moving rats. J Pharmacol Exp Ther 1988; 244:1067-1080.
38. Imperato A, Di Chiara G. Preferential stimulation of dopamine-release in the accumbens of freely moving rats by ethanol. J Pharmacol Exp Ther 1986; 239: 219-228.
39. Imperato A, Mulas A, Di Chiara G. Nicotine preferentially stimulates dopamine release in the limbic system of freely moving rats. Eur J Pharmacol 1986; 132:337-338.
40. Carboni E, Imperato A, Perezzani L et al. Amphetamine, cocaine, phencyclidine and nomifensine increase extracellular dopamine concentrations preferentially in the nucleus accumbens of freely moving rats. Neurosci 1989; 28:653-661.
41. Radhakishun FS, van Ree JM, Westerink BHC. Scheduled eating increases dopamine release in the nucleus accumbens of food-deprived rats as assessed with on-line brain dialysis. Neurosci Lett 1988; 35: 351-356.
42. Hernandez L, Hoebel BG. Feeding and hypothalamic stimulation increase dopamine turnover in the accumbens. Physiology and Behavior 1988; 44:599-608.
43. Hernandez L, Hoebel BG. Food reward and cocaine increase extracellular dopamine in the nucleus accumbens as measured by microdialysis. Life Sci 1988; 42:1705-1712.
44. Evans KR, Eikelboom R. Feeding induced by ventricular bromocriptine and amphetamine: a possible excitatory role for dopamine in eating behavior. Behav Neurosci 1987; 101:591-593.
45. Evans KR, Vaccarino FJ. Effects of d- and l-amphetamine on food intake: evidence for a dopaminergic substrate. Pharmacol Biochem Behav 1987; 27:649-652.

46. Evans KR, Vaccarino FJ. amphetamine- and morphine-induced feeding: evidence for involvement of reward mechanisms. Neurosci Biobehav Rev 1990; 14:9-22.
47. Grinker JA, Drewnowski A, Enns M et al. The effects of d-amphetamine and fenfluramine on feeding patterns and activity of obese and lean Zucker rats. Pharmacol Biochem Behav 1980; 12:265-275.
48. Holtzman SG. Behavioral effects of separate and combined administration of naloxone and d-amphetamine. J Pharmacol Exp Ther 1974; 189:51-60.
49. Winn P, Williams SF, Herberg LJ. Feeding stimulated by very low doses of d-amphetamine administered systemically or by microinjection into the striatum. Psychopharmacol 1982; 78:336-341.
50. Colle LM, Wise RA. Facilitory and inhibitory effects of nucleus accumbens amphetamine on feeding. Ann NY Acad of Sci 1988; 537:491-492.
51. Evans KR, Vaccarino FJ. Intra-nucleus accumbens amphetamine: dose-dependent effects on food intake. Pharmacol Biochem Behav 1986; 25:1149-1151.
52. Piazza PV, Deminiere JM, Le Moal M et al. Factors that predict individual vulnerability to amphetamine self-administration. Science 1989; 245:1511-1513.
53. Hooks MS, Jones GH, Smith AD et al. Response to novelty predicts the locomotor and nucleus accumbens dopamine response to cocaine. Synapse 1991; 9:121-128.
54. Piazza PV, Rouge-Pont F, Deminiere JM et al. Dopaminergic activity is reduced in the prefrontal cortex and increased in the nucleus accumbens of rats predisposed to develop amphetamine self-administration. Brain Res 1991; 567:169-174.
55. Hooks MS, Colvin AC, Juncos JL et al. Individual differences in basal and cocaine stimulated extracellular dopamine in the nucleus accumbens using quantitative microdialysis. Brain Res 1992; 587:306-312.
56. Robinson TE, Becker JB. Enduring changes in brain and behavior produced by chronic amphetamine administration: a review and evaluation of animal models of amphetamine psychosis. Brain Res Rev 1986; 11:157-198.
57. Robinson TE, Jurson PA, Bennett JA et al. Persistent sensitization of dopamine neurotransmission in ventral striatum (nucleus accumbens) produced by prior experience with (+)-amphetamine: a microdialysis study in freely moving rats. Brain Res 1988; 462:211-222.
58. Stewart J, Vezina P. Conditioning and behavioral sensitization. In: Kalivas PW,.Barnes CD, eds. Sensitization in the Nervous System. Caldwell, N.J.: Telford Press, 1988:207-224.
59. Weiss SRB, Post RM, Pert A et al. Context-dependent cocaine sensitization: differential effect of haloperidol on development versus expression. Pharmacol Biochem Behav 1989; 34:655-661.

60. Hooks MS, Jones GH, Neill DB et al. Individual differences in amphetamine sensitization: dose-dependent effects. Pharmacol Biochem Behav 1992; 41:203-210.
61. Sills TL, Vaccarino FJ. Individual differences in sugar intake predict the locomotor response to acute and repeated amphetamine administration. Psychopharmacol 1994; 116:1-8.
62. Piazza PV, Rogue-Pont F, Deminiere JM et al. Individual vulnerability to amphetamine self-administration is correlated with dopaminergic activity in frontal cortex and nucleus accumbens. Soc Neurosci Abstr 1990;16.
63. Corrigal W, Vaccarino FJ. Antagonist treatment in Nucleus Accumbens or Periaqueductal grey affects heroin self-administration. Pharmacol Biochem Behav 1988; 30:443-450.
64. Glick SD, Merski C, Steindorf S et al. Neurochemical predisposition to self-administer morphine in rats. Brain Res 1992; 578: 215-220.
65. Sills TL, Vaccarino FJ. Facilitation and inhibition of feeding by a single dose of amphetamine: relationship to baseline intake and accumbens cholecystokinin. Psychopharm 1991; 105:329-334.
66. Sills TL, Vaccarino FJ. Baseline-dependent feeding responses to amphetamine: effects of intra-accumbens alpha-flupenthixol administration. Soc Neurosci Abstr 1991; 17:327.
67. Sills TL, Baird JP. Vaccarino FJ. Individual differences in the feeding effects of amphetamine: role of nucleus accumbens dopamine and circadian factors. Psychopharmacol (in press).
68. Carr GD, White NM. Contributions of dopamine terminal areas to amphetamine-induced anorexia and adipsia. Pharmacol Biochem Behav 1986; 25:1-6.
69. Mitchell JB, Gratton A. Opioid modulation and sensitization of dopamine release elicited by sexually relevant stimuli: A high speed chronoamperometric study in freely behaving rats. Brain Res 1991; 551:20-27.
70. Mitchell JB, Stewart J. Effects of castration, steroid replacement, and sexual experience on mesolimbic dopamine and sexual behaviors in the male rat. Brain Res 1989; 491:116-127.
71. Mitchell JB, Stewart J. Facilitation of sexual behaviors in the male rat associate with intra-VTA injections of opiates. Pharmacol Biochem Behav 1990; 35:643-650.
72. Vandereycken W. The addiction model in eating disorders: Some critical remarks and a selected bibliography. Int J Eating Disorders 1990; 9:95-101.
73. National Institute of Drug Abuse. National household survey on drug abuse. Rockville, M.D., 1982.
74. Gilles C, Lotstra F, Vanderhaegen JJ. CCK nerve terminals in the rat striatal and limbic areas partly originate in the brain stem and partly originate in telencephalic structures. Life Sci 1983; 32:1683-1690.

75. Hökfelt T, Skirboll L, Rehfeld JF et al. A subpopulation of mesencephalic dopamine neurons projecting to limbic areas contain a cholecystokinin-like peptide: evidence from immunohistochemistry combined with tracing. Neurosci 1980; 5:2093-2124.
76. Seroogy KM, Fallon JH. Forebrain projections from cholecystokinin-immunoreactive neurons in the rat brain. J Comp Neurol 1989; 279:415-435.
77. Studler JM, Rebaud M, Tram G et al. Distinct properties of CCK-8 and mixed dopamine/CCK-8 neurons innervating the nucleus accumbens. In: eds Vanderhaegen JJ, Crawley JN. Ann NY Acad Sci 1985; 448:306-314.
78. Studler JM, Reibaud M, Tramu G et al. Pharmacological study on the mixed CCK-8/DA meso-nucleus accumbens pathway: evidence for the existence of storage sites containing the two transmitters. Brain Res 1984; 298:91-97.
79. Studler JM, Simon H, Ceselin F. Biochemical investigation of the localization of CCK octapeptide in dopaminergic neurons originating from the ventral tegmental area of the rat. Neuropeptides 1981; 2:131-139.
80. Beinfeld MC, Meyer DK, Eskay RL et al. The distribution of CCK immunoreactivity in the CNS of the rat as determined by radioimmunoassay. Brain Res 1981; 212:51-57.
81. Golterman NR, Stengaard-Pedersen H, Rehfeld JF et al. Newly synthesized cholecystokinin in subcellular fractions of rat brain. J Neurochem 1981; 36:859-865.
82. Emson PC, Lee CM, Rehfeld JF. Cholecystokinin octapeptide: vesicular localization and calcium dependent release from rat brain in vitro. Life Sci 1980; 26:2157-2163.
83. Moran TH, Robinson PH, Goldrich MS et al. Two brain cholecystokinin receptors: implications for behavioral actions. Brain Res 1986; 362:175-179.
84. Crawley JN, Hommer DW, Skirboll LR. Topographical analysis of nucleus accumbens sites at which cholecystokinin potentiates dopamine-induced hyperlocomotion in the rat. Brain Res 1985; 335:337-341.
85. Crawley JN, Stivers JA, Blumstein LK et al. Cholecystokinin potentiates dopamine-mediated behaviors: evidence for modulation specific to a site of coexistence. J Neurosci 1985; 5:1972-1983.
86. van Ree JM, Gaffori O, De Wied D. In rats, the behavioral profile of CCK-8 related peptides resembles that of antipsychotic agents. Eur J Pharmacol 1983; 93:63-78.
87. Weiss F, Tanzer DJ, Ettenberg A. Opposite actions of CCK-8 on amphetamine-induced hyperlocomotion and stereotypy following intracerebroventricular and intra-accumbens injections in rats. Pharmacol Biochem Behav 1986; 30:309-317.
88. Vaccarino FJ, Rankin J. Nucleus accumbens cholecystokinin (CCK) can either attenuate or potentiate amphetamine-induced locomotor activity: evidence for rostral-caudal differences in accumbens CCK function. Behav Neurosci 1989; 103:831-836.

89. Blaha CD, Phillips AG, Lane RF. Reversal by cholecystokinin of apomorphine-induced inhibition of dopamine release in nucleus accumbens of the rat. Reg Peptides 1987; 17: 301-310.
90. Crawley JN. Subtype-selective cholecystokinin receptor antagonists block cholecystokinin modulation of dopamine-mediated behaviors in the rat mesolimbic pathway. J Neurosci 1992; 12:3380-3391.
91. Crawley JN, Stivers JA, Hommer DW et al. Antagonism of central and peripheral behavioral actions of cholecystokinin octapeptide. J Pharmacol Exp Ther 1986; 236:320-330.
92. Marshall FH, Barnes S, Hughes J et al. Cholecystokinin modulates the release of dopamine from the anterior and posterior nucleus accumbens by two different mechanisms. J Neurochem 1991; 56:917-922.
93. Heidbreder C, Gewiss M, De Mot B et al. Balance of glutamate and dopamine in the nucleus accumbens modulates self-stimulation behavior after injection of cholecystokinin and neurotensin in the rat brain. Peptides 1992; 13:441-449.
94. Phillips GD, Westergreen A-LM, Robbins TE et al. Functional synergism within the ventral striatum between cholecystokinin and dopamine in the control over behaviour by conditioned reinforcers. J Psychopharmacol 1992; 6:115.
95. Vaccarino FJ, Koob GF. Microinjections of nanogram amounts of sulfated cholecystokinin octapeptide into the rat nucleus accumbens attenuates brain stimulation reward. Neurosci Lett 1984; 52:61-66.
96. Vaccarino FJ, Weiss F, Koob GF. Neuroleptic-likes effects of intra-accumbens CCK on iv cocaine self-administration. Soc Neurosci Abstracts 1989; 15:1097.
97. Ettenberg A, Pettit HO, Bloom FE et al. Heroin and cocaine intravenous self-administration in rats: Mediation by separate neural systems. Psychopharmacol 1982; 78:204-209.
98. De Witte P, Heidbreder C, Roques B et al. Opposite effects of cholecystokinin octapeptide (CCK-8) and tetrapeptide (CCK-4) after injection into the caudal part of the nucleus accumbens or into its rostral part and the cerebral ventricles. Neurochem Int 1987; 10:473-479.
99. Vaccarino FJ, Vaccarino AL. Antagonism of cholecystokinin function in the rostral and caudal nucleus accumbens: differential effects on brain stimulation reward. Neurosci Lett 1989; 97:151-156.
100. Higgins GA, Nguyen P, Sellers EM. Morphine place conditioning is differentially affected by $CCK_A$ and $CCK_B$ receptor antagonists. Brain Res 1992; 572:208-215.
101. Josselyn SA, Vaccarino FJ. Blockade of CCK-B receptors potentiates responding for conditioned rewards. Soc Neurosci Abstr 1994; 20:380.
102. Higgins GA, Tomkins DM, Sills TL et al. Effects of CCK-A and CCK-B antagonists on amphetamine-induced behaviors in rats. Soc Neurosci Abstr 1992; 18:899.

103. Higgins GA, Sills TL, Tomkins DM et al. Evidence for the contribution of CCK-B receptor mechanisms to individual differences in amphetamine-induced locomotion. Pharmacol Biochem Behav 1994; 48:1019-1024.
104. Rossetti ZL, Melis F, Carboni S et al. Dramatic depletion of mesolimbic extracellular dopamine after withdrawal from morphine, alcohol or cocaine: a common neurochemical substrate for drug dependence. In: Kalivas PW, Samson HH, eds. The Neurobiology of Drug and Alcohol Addiction. New York: Ann NY Acad Sci, 1992:513-521.
105. Koob GF, Stinus L, Le Moal M et al. Opponent process theory of motivation neurobiological evidence from studies of opiate dependence. Neurosci Biobehav Rev 1989; 13:135-140.

CHAPTER 8

# Opioid and CCK Systems in Anxiety and Reward

Valérie Daugé and Bernard P. Roques

## INTRODUCTION

CCK was first recognized as a gastrointestinal hormone, mediating digestive function and feeding behavior. Nevertheless numerous physiological processes such as respiratory functions[1] and cardiovascular tonus,[2] vigilance states,[3] memory processes,[4-6] nociception[7-11] and emotional and/or motivational states[12-14] were also shown to be modulated by CCK. The sulfated octapeptide fragment of cholecystokinin (CCK-8) which interacts with the same affinity with CCK-A and CCK-B receptors (Table 8.1) is the most abundant form found in the brain. Low amounts of shorter fragments such as the C-terminal tetrapeptide CCK-4, which exhibits a better selectivity for CCK-B than for CCK-A receptors (Table 8.1) were also shown to occur in the brain.[15]

The distribution of the two types of CCK receptors in the organism is quite different. The A type is abundant in the peripheral system but is also found in a few discrete regions of the rat brain i.e., area postrema, interpeduncular nucleus, posterior hypothalamus, posterior nucleus accumbens, amygdala and nucleus tractus solitarius (NTS).[16,17] In contrast the B type is the predominant

*Cholecystokinin and Anxiety: from Neuron to Behavior*, edited by Jacques Bradwejn and Eero Vasar. 

*Table 8.1. Affinities of CCK analogs on rat pancreas (CCK-A) and rat brain (CCK-B) binding sites*

| Compounds | | CCK-A $K_i$ nM $^3$HpCCK-8 | CCK-B $K_i$ nM $^3$HpBC 264 | CCK-A / CCK-B |
|---|---|---|---|---|
| CCK-B agonists | | | | |
| **CCK-8** | Asp-Tyr($SO_3H$)-Met-Gly-Trp-Met-Asp-Phe$NH_2$ | 0.87 ± 6 | 1.01 ± 0.05 | 0.86 |
| **CCK-4** | Trp-Met-Asp-Phe$NH_2$ | 5113 ± 02 | 55.7 ± 10.2 | 92 |
| **BC 197** | Boc-(D).Asp-Tyr($SO_3H$)-Nle-(D).Lys-Trp-Nle-Asp-Phe$NH_2$ | 1845 ± 402 | 160 ± 52 | 11.5 |
| **BC 264** | Boc-Tyr($SO_3H$)-gNle-mGly-Trp-(NMe)Nle-Asp-Phe$NH_2$ | 355 ± 98 | 0.39 ± 8 | 910 |
| CCK-B antagonists | | | | |
| **RB 211** | N-(2-ademantyloxycarbonyl)-(D)-α-methyl-tryptophanyl-[N-{2-(4-chlorophenyl)ethyl}]glycine. | 1232 ± 130 | 20.9 ± 4.5 | 59 |
| **AB 36**[1] | | 2210 ± 432 | 15 ± 1 | 147 |
| **CI-988** | [R-(R*,R*)]-4-[[2-[[3-((1H-Indol-3-yl)-2-methyl-1-oxo-2-[[(tricyclo[3.3.1.1.",7]dec-2-yloxy) carbonyl]amino]propyl]-amino]-1-phenylethyl]amino]-4-oxobutanoic acid. | 2382 ± 514 | 6.1 ± 0.5 | 390 |
| **L-365,260** | (3R)-N-(2,3-dihydro-1-methyl-2-oxo-5-phenyl-1H-1,4-benzodiazepin-3-yl)-N′-(3-methylphenyl)urea. | 1170 ± 330 | 10.9 ± 2 | 107 |
| CCK-A antagonist | | | | |
| **L-364,718 or MK 329** | (3S)-N-(2,3-dihydro-1-methyl-2-oxo-5-phenyl-1H-1,4-benzodiazepin-3-yl)-1H-indole-2-carboxamide. | 1.55 ± 0.15 | 549 ± 69 | 0.0028 |

[1] : Values obtained on guinea-pig pancreas and brain.

form found in the brain, particularly in the cerebral cortex and in limbic structures such as the olfactory tubercles, hippocampus, amygdala and anterior nucleus accumbens (N.Acc),[18] raising interesting questions as to its possible role in psychiatric disorders including schizophrenia, depressive syndromes and anxiety. The recent purification and molecular cloning of the CCK receptors from rat, guinea-pig and human pancreas (CCK-A) and brain (CCK-B) have shown that these two distinct receptors contain seven putative transmembrane domains, suggesting that they belong to the guanine nucleotide-binding regulatory protein-coupled receptor superfamily.[19,20] Genomic and cDNA library hybridization as well as Northern and Southern hybridization studies among rat, guinea-pig and human species identifies only two members of the CCK receptor family, CCK-A and CCK-B. The stomach gastrin receptor and the brain CCK-B receptor appear to be identical.[20] The rat CCK-B receptor gene contains several exons and different forms of mRNAs were shown discretely distributed in various rat brain structures.[21] Nevertheless, differences in tissue and species-specific posttranslational processing, receptor coupling and associated membrane protein and lipid heterogeneity may account among some other factors, for the phenotypic expression of more receptor subtypes than molecular studies would predict.[22,23]

Extensive pharmacological studies have shown that CCK-4 and the related molecule pentagastrin, can produce panic attacks after iv administration in humans.[24,25] Indeed, CCK-4 was found to induce various symptoms-anxiety, apprehension, tension and fear, in addition to somatic and cognitive symptoms characteristic of panic attacks. These effects were suppressed by the CCK-B antagonist, L-365,260.[26] Interestingly, patients with panic disorders were found to be more sensitive to the effects of CCK-4 than healthy volunteers.[27] Furthermore, a reduction in the CCK concentration was detected in the cerebrospinal fluid of these patients.[28] Therefore, the possibility that CCK plays a role in the neurobiology of anxiety and panic attacks was investigated by several laboratories.[26] The contribution of CCK-A or/and CCK-B receptors in the pathophysiology role of CCK-8 was therefore of particular great interest.

However, one of the major problems encountered with neuropeptides is their relatively low tonic release contrasting with the high concentrations of neurotransmitters permanently secreted from neurons in the central nervous system. This phenomenon explains that antagonists, such as naloxone in the case of the opioid peptides system have practically no effects in humans and animals. A physiological way to overcome this problem is either to increase the extracellular concentration of neuropeptides by inhibiting their peptidase-induced inactivation as illustrated in the case of enkephalins (for review see ref. 29). Another possibility could be to trigger the release of the studied neuropeptide by adapted stimuli. However, this requires knowledge of the pharmacological responses associated with the selective stimulation of the peptide receptor which is exactly what we are searching for.

Therefore, the simplest way would be to use highly potent, selective and systemically active agonists to overstimulate a particular peptide receptor. In conjunction with antagonists these compounds would allow the putative clinical interest of activating or blocking a peptidergic pathway to be easily investigated.

## DESIGN OF CCK-8-DERIVED PEPTIDOMIMETICS SELECTIVE FOR CCK-B RECEPTOR

Different strategies have been followed to design potent and selective agonists and antagonists of CCK-B receptors. In spite of its intrinsic flexibility, CCK-8 was found by NMR to exist preferentially under folded form in aqueous solution,[30] with a proximity between Asp1 and Gly4. This property was used to synthesize cyclic peptides through amide bond formation between Asp1 or α, β Glu1 and Lys4 side chains.[31] Compounds such as BC 254 and BC 197 were found highly potent and selective CCK-B agonists, but their affinities for the CCK-B sites was found species dependent suggesting the presence of CCK-B receptor subsites corresponding probably to different modes of G-coupling of the receptor.[23]

Another approach towards CCK-B agonists was to protect CCK-8 from degrading enzymes such as aminopeptidase A (APA) and a thiol/serine protease cleaving this peptide at the Met-Gly bond. This led to BC 264, a highly potent CCK-B agonist which exhibits about the same affinity ($K_D$ = 0.1 to 0.5 nM) in all species (guinea-pig, rat, mouse, monkey, human, transfected cell) and was at that time the only systemically active B agonist.[32-34] [$^3$H]pBC 264 was shown to enter the brain in which it is not significantly metabolized.[35] Peptidomimetics such as AB 36 endowed with CCK-

B agonist properties have been obtained by partial cyclisation of CCK-4 derived agonists.[36-37] The design of these molecules resulted from NMR and molecular modeling studies[38] (Fig. 8.1). The pharmacological profile of all these compounds was established by using peripheral tests and CHO-cells in which the rat CCK-B receptor has been transfected and found to be coupled to inositol phosphate system (Jagerschmidt et al, in preparation) (Table 8.1).

## EVIDENCE FOR THE INVOLVEMENT OF CCK IN "ANXIETY"/FEAR IN RODENTS AND MONKEYS

Although, there is no data in rodents corresponding to panic attacks in humans due to the lack of a specific animal model of this disease, several studies indicate that CCK may act as an endogenous anxiogenic peptide. Thus, various CCK-B antagonists [L-365,260, CI-988 (Table 8.1)] were able to produce anxiolytic effects in several models of fear and/or anxiety in rodents, such as the black and white box in mice[14] the elevated plus maze in mice and rats,[14,39] an ethological model for measuring behavioral responses to

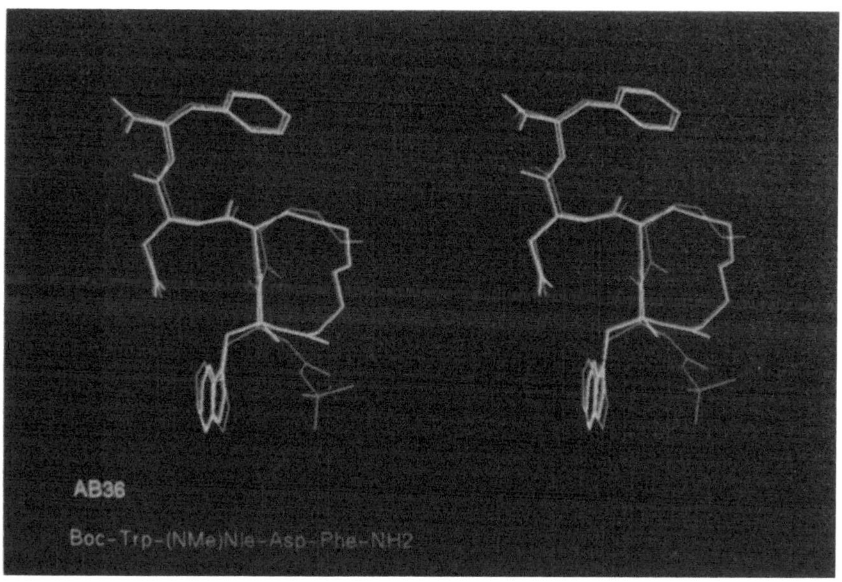


*Fig. 8.1. Superposition of the biologically active conformations deduced from NMR and template forcing calculations of the CCK-4-derived selective and potent CCK-B agonist Boc-Trp-(NMe)Nle-Asp-Phe-$NH_2$ and its cyclic peptidomimetic AB 36. The cyclisation was achieved by using the amino terminus and the side chain of a linear derivative of Nle.*

predation,[40] social interactions in rats and in a marmoset "human threat" model[41] and conflict (alternations of footshock punished and unpunished periods) tests in rats.[14] The CCK-B antagonists were also efficient but with a lesser extent to produce anxiolytic-like effects in punished exploratory behavior.[42] Moreover, the involvement of CCK-B receptors, in these tests, was emphasized by the anxiogenic-like effects produced by the selective CCK-B agonists, (BC 264 or BC 197) and their suppression by CCK-B antagonists (CI-988, L-365,260) in the black and white box test in mice.[43] Furthermore a new selective CCK-B antagonist, RB211, synthesized in the laboratory[44] produced anxiolytic effects after ip injection in mice, supporting a physiological regulation of emotional states by CCK.[43] Boc-CCK-4 was also found to produce anxiety-like behavior in several ethologically based animal models of anxiety in rats.[45]

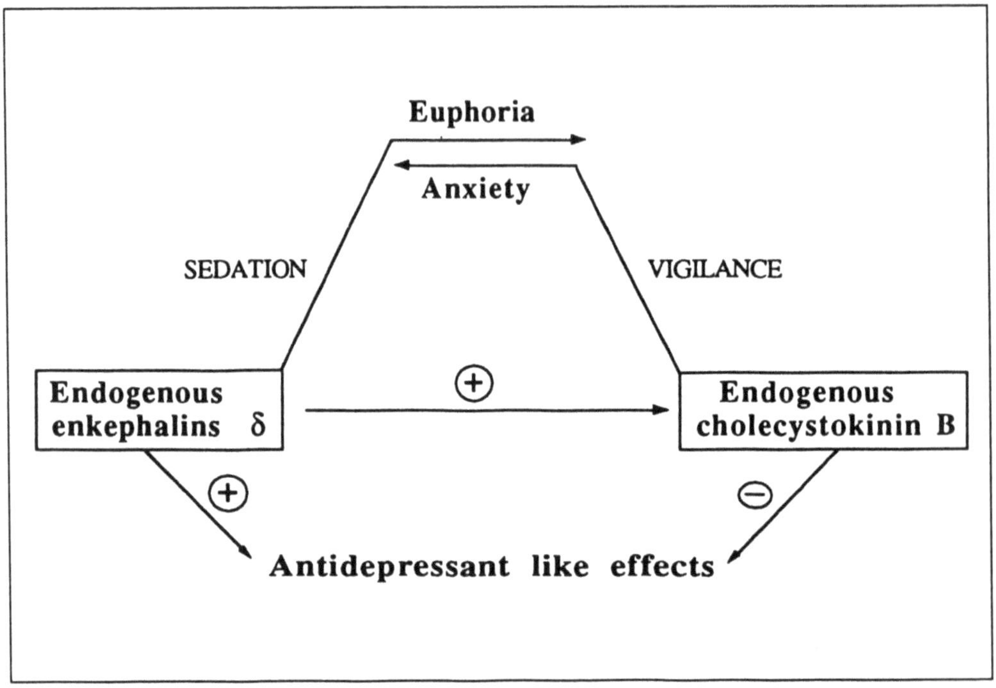


*Scheme 8.1. Schematic representation of the relationship between CCKergic and enkephalinergic systems in behavioral control.*
*Activation of δ opioid receptors by the endogenous enkephalins or blockade of the activity of CCKergic system by CCK-B antagonists induced antidepressant-like effects, in the conditioned suppression of motility and forced swim (Porsolt) tests in mice. The effects of endogenous enkephalins were potentiated by CCK-B antagonists indicating an opposite physiological interaction between these two neuropeptides. Enkephalins are involved in rewarding behaviors, sedation and euphoria while CCK counteract these effects by increasing vigilance and anxiety.*

In contrast to CCK-B, CCK-A receptor stimulation induced sedative effects when a CCK-4-derived CCK-A agonist, A5b,Boc-Tyr-Lys(CO-NH-o-tol)-Asp-PheNH$_2$[46] was ip administered in the black and white box test and in Animex meter studies.[47] Nevertheless, CCK-A receptor activation has been claimed by some authors to produce anxiogenic effects[40] (reviewed in ref. 48).

The often observed lack of dose-response curves following administration of nonselective compounds in behavioral studies could result from successive activation of CCK-B and CCK-A receptors (reviewed in ref. 48). Furthermore, the effects of CCK compounds could vary considerably in regard to the studied animals due to differences in the distribution and binding characteristics of CCK receptor types among species[20,49,50] or subtypes.[22,51] Indeed recently, the effects of selective CCK-B agonists, BC 264 and BC 197 and the non-selective CCK agonist BDNL were investigated in the elevated plus maze in rats. BDNL and BC 197 induced anxiogenic-

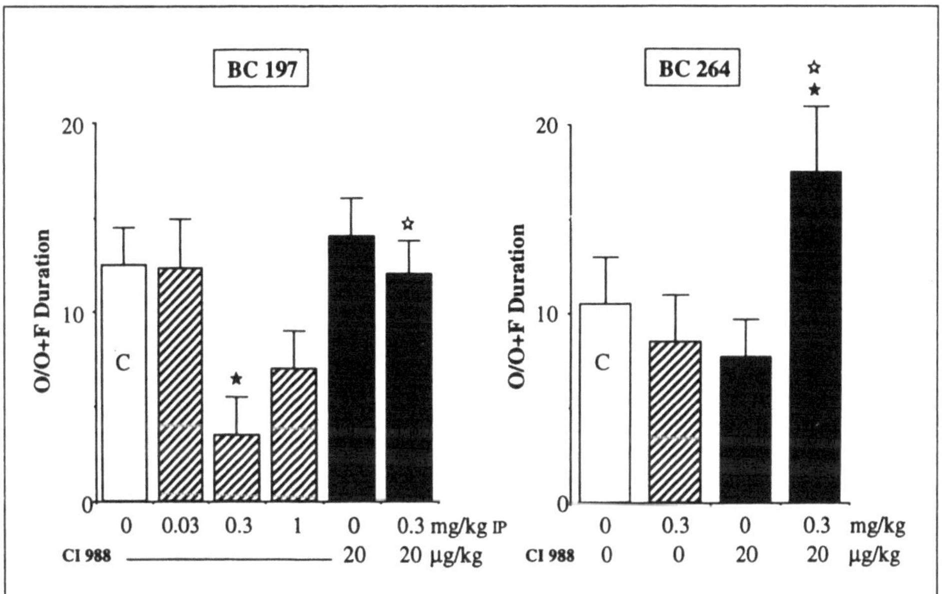


*Fig. 8.2. The CCK-B agonist BC 197, 30 min after i.p. injection, induced an anxiogenic-like effect measured in the elevated plus maze in rat (decreased percentage of the time spent in open arms). This effect was suppressed by the ip injection of the CCK-B antagonist CI-988 45 min before the test. The CCK-B agonist BC 264 30 min after ip injection, did not modify the time spent in open arms, whereas ip administration of CI-988 15 min before BC 264 injection increased it. These data suggest the existence of CCK-B receptor subtypes. CI-988 might selectively block one of the CCK-B receptor subsites, allowing BC 264 to interact with the other subsite. Newman-Keuls test, ★$P < 0.05$ vs control. ☆$P < 0.05$ vs CCK-B agonist.*

like effects, in contrast to BC 264, which was devoid of effect. The behavioral responses induced by BDNL and BC 197 were suppressed by CI-988 but not by L-365,260, suggesting some subsite selectivity for these two CCK-B antagonists[52] (Fig. 8.2). According to these findings, competition experiments performed with [$^3$H]pBC 264 and brain membranes of guinea-pig, mouse and rat were significantly better fitted when analyzed by a two site model than by a one site model with BC 197 but not with BC 264.[52] In the black and white box test in mice, BC 197 was found to induce anxiogenic-like effects at doses as low as 1 μg/kg ip (Fig. 8.3) while an approximately 300 times higher dose of BC 264 was required to observe the same responses.

Ervin et al have recently studied the anxiogenic effects of CCK agonists in a nonhuman primate model.[33] CCK-4 given iv to African green monkeys has strong and dose-related effects on behaviors thought to reflect anxiety and panic. Interestingly, the CCK-B agonist BC 264, which is known to penetrate the brain efficiently,[33] also produces these behavioral responses, but the profile of behavior is somewhat different since at low concentration, hypervigilance and stereotypy were prominent. There is no attenuation of the effects produced by CCK-4 or BC 264 when a CCK-B antagonist, which does not effectively cross the blood-brain barrier,

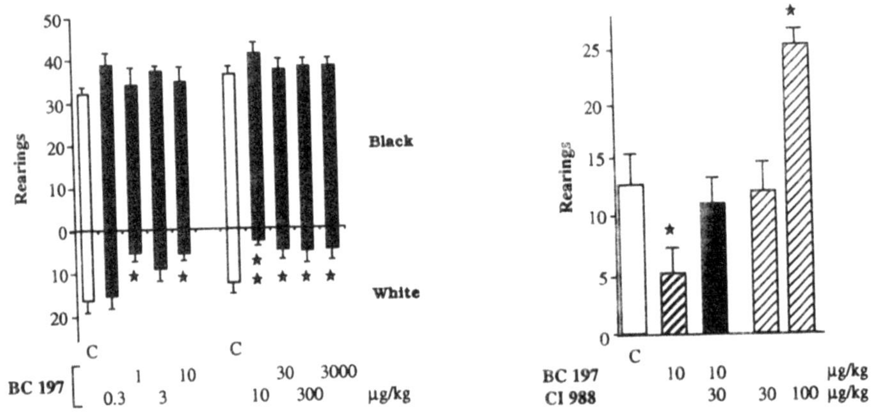


*Fig. 8.3. Anxiogenic-like effect, (decrease the number of rearings in the white box) measured in the black and white box test in mice of the CCK-B agonist, BC 197 30 min after i.p. injection. The CCK-B antagonist i.p. injected at the dose of 100 μg/kg 45 min before the test, induced an anxiolytic-like effect (increase of rearings). When it was injected before BC 197 at the dose of 30 μg/kg, CI-988 completely prevented the effects of BC 197. * P < 0.05 ; ** P < 0.01, Newman-Keuls test.*

RP71483, is injected.[53] In contrast, CCK-4 effects are potently and dose-responsively blocked by various lipophilic CCK-B antagonists assumed to enter the CNS [LY262691, LY247348, CI-988 (Table 8.1)], but not by the CCK-A antagonist (LY219057). These data are therefore consistent with the hypothesis that the main behavioral effects of CCK-4 are mediated centrally, although the participation of peripheral receptors cannot be completely excluded. In agreement with the results found in monkeys preliminary experiments indicate that BC 264 could produce an increase of behavioral arousal after ip administration in rats at very low dose (~μg/kg). It also induced an increase of the extracellular level of dopamine and its metabolites (DOPAC and HVA) in the nucleus accumbens using microdialysis technique in awake and freely moving rats (Ladurelle et al, in preparation).

## ANATOMICAL SUBSTRATE OF CCK-INDUCING ANXIOGENIC-LIKE EFFECTS

There is some evidence for the participation of particular brain structures in the behavioral expression of the anxiogenic-like effects and the associated neurovegetative and neuroendocrine changes induced by CCK agonists. The solitary complex (SC) which is composed of dorsal motor nucleus of the vagus (DVMN) and the nucleus tractus solitarius (NTS), plays a major role in translating peripheral information to the brain. This was clearly demonstrated in the case of the satiety reflex which involves initial peripheral CCK-A receptor activation.[54-57,72] The NTS was postulated to directly or indirectly participate in some components of panic attacks[58] triggered by stimulation of CCK-B binding sites. Indeed CCK-B receptor stimulation by BC 264 induces postsynaptic excitation of neurons involved in cardiorespiratory control and in emotional responses through activation of direct or indirect ascending pathways projecting to the N.Acc,[59] amygdala, periaqueductal grey matter and locus ceruleus (reviewed in ref. 60). The effect of BC 264 was reversed by CCK-B antagonists L-365,260 and CI-988 and could involve a presynaptic inhibition of inhibitory GABAergic interneurons. Interestingly, CCK-B antagonists alone produced a reduction in basal firing of neurons of the SC suggesting a tonic control by CCK-8.[58]

Other brain regions linked or not to the NTS seem to be involved in CCK-induced anxiogenic effects. Thus, CCK-8 has been shown to activate rat hippocampal neurons, an effect suppressed by activation of the neuronal type of benzodiazepine receptor.[61] The authors have suggested that this electrophysiological link between CCK and benzodiazepine might be related to the anxiolytic properties of benzodiazepine. Moreover local administration of CCK-8 in the dorsal periaqueductal grey matter which contains neurons where CCK and enkephalins are colocalized, induced a selective decrease in exploration of the aversive part (open arms) of the elevated plus maze in rats.[62] Besides, microinjection of CCK-8 and unsulfated CCK-8 into the central amygdaloid nucleus delayed the extinction of active avoidance and increased the retention of passive avoidance behavior which have been possibly related to an enhancement of behavioral arousal and/or fear-motivation of the animals.[63]

More complex results have been described after local injection of CCK-8 into the N.Acc, that may be due to the anatomical and biochemical heterogeneity of the CCKergic system in this brain structure. While neuroleptic-like effects have been found with CCK-8,[63,64] a hypoexploration in the four hole box and in the open-field (only observed in a novel environment) and an anxiogenic-like effect in the elevated plus maze were observed after stimulation of CCK-A receptors in the postero-median part of the N.Acc, but not into the anterior part.[12,13,65] Interestingly this part of the nucleus receives DA-CCK colocalized fibers issuing from the ventral tegmental area[66] but also CCK fibers from the NTS.[59] In agreement with this neuroanatomical organization, CCK-8 induced effects were suppressed by 6-hydroxydopamine lesions of the DA mesoaccumbens fibers and by dopamine $D_2$ but not $D_1$ antagonists.[67] On the other hand, the anterior part of the N. Acc (especially lateral part) receives CCK fibers from the prefrontal cortex.[68] The local injection of BC 264, in this part of the N. Acc, reduced the spontaneous alternation behavior of rats.[69]

Moreover, data obtained using microdialysis technique in awake rats showed that the stimulation of CCK-B receptors by BC 264, locally perfused in the rostral part of the N. Acc, decreased the extracellular level of dopamine while CCK-8 in the posterior part of the N. Acc increased it in agreement with experiments performed on N. Acc slices.[70-71] These data suggest that the CCK modulation of DA involves two different mechanisms in the rostral and the caudal regions of the N. Acc which could be related to the heterogeneity of responses obtained in CCK behavioral studies.[13,69,55] The posteromedian part of the N. Acc (shell region) is of particular interest because it is connected with the mesencephalic region by CCK-DA colocalized fibers[66,73] and it projects to a subterritory of ventral pallidum which in turn innervated the mediodorsal thalamic nucleus and then the prefrontal cortex. Due to these connections and pharmacological data, the shell N. Acc is considered as the limbic part of this structure.[74,75]

After local injection of CCK-8, in the posterior part of N. Acc, hypoexploration and anxiogenic-like responses were obtained only when the rats were stimulated by a novel environment.[12,13] Therefore, experiments were performed to investigate the influence of a novel environment on DA release. Results showed that the transfer of rat to a novel environment slightly increased the DA release. Furthermore, CCK-8 produced a longer effect in rats transferred to a novel environment since the DA efflux was still increased 80 min after the removal of CCK-8[76] (Fig. 8.4). Moreover, an anxiogenic-like effect of CCK-8 was found in the elevated plus maze, only in rats submitted to a novel environment. This indicates that the intensity of the CCK-8 effects on extracellular DA levels and on anxiety-like responses seems to depend on the activity of these neurons previous to CCK-8 treatment.[76] This is in line with the observation that particular experimental conditions

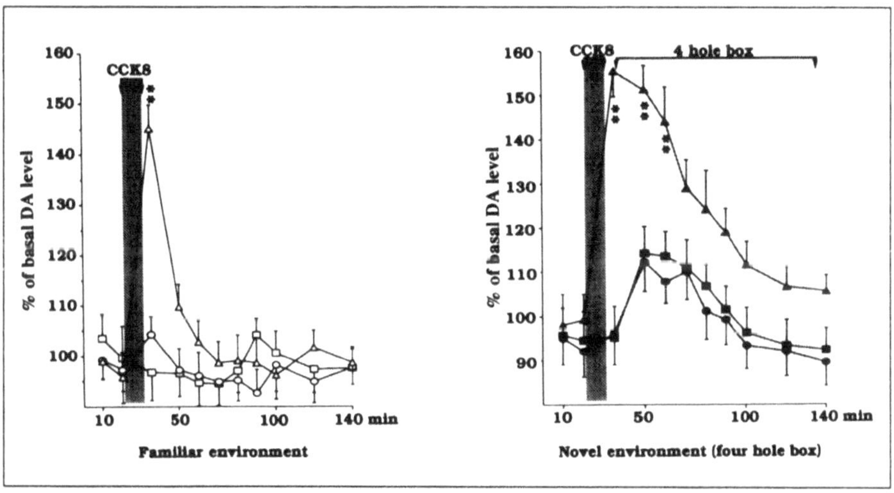


*Fig. 8.4. Perfusion of 25 pmol (△, ▲) but not 1 fmol (○, ●) of CCK-8 in the posterior nucleus accumbens increased for 10 min, the extracellular level of dopamine, when rats were acclimated to the environment. Same perfusion of CCK-8 prolonged for 80 min, the increase of extracellular level of dopamine when rats were submitted to a novel environment (the four hole box). The experiments were performed using the microdialysis technique. The quantification of dopamine was done by HPLC with coulometric detection.*

suggest that the clinical use of CCK-B antagonists, restricted until now to inhibition of panic attacks, could be extended to the treatment of depressive syndromes.

## CONCLUSION

The two peptidergic systems using enkephalins or CCK as endogenous effectors seem to be activated in several stressful and conflicting situations in humans, monkeys and rodents. The phasic release of endogenous CCK-8 could lead to various behavioral responses such as nociception and anxiety through interaction with CCK-B receptors. The enkephalinergic system appears to produce opposite effects through activation of $\mu$ and $\delta$ receptors respectively. In addition, our recent results indicate that the stimulation of CCK-B receptors by BC 264 at low doses could enhance vigilance and/or attention of rats and monkeys. This suggests that activation of CCK-B receptor by endogenous CCK is required to maintain a cognitive and emotional level required for being able to detect unusual or possible dangerous events in the environment. While the physiological circumstances of their release remain speculative, studies performed in our laboratory with inhibitors of peptide catabolism and antagonists show that CCK and enkephalins participate in opposite manner to the coordinated control of some adaptative behaviors.

## REFERENCES

1. Morin-Surun MP, De Marchi J, Vanderhaeghen JJ et al. Inhibitory effect of cholecystokinin octapeptide on neurons in the nucleus tractus solitarius. Brain Res. 1983; 265:333-338.
2. Koyama S, Fujita T, Shibamoto T et al. Contribution of baroreceptor reflexes to blood pressure and sympathetic responses to cholecystokinin and vasoactive intestinal peptide in anaesthesized dogs. Eur J Pharmacol 1990; 175:245-251.
3. Saint-Hilaire Z, Roques BP, Nicolaidis S. Effect of a highly selective central CCK-B receptor agonist: BC 264 on rat sleep. Pharmacol Biochem Behav 1991; 38:545-548.
4. Itoh S, Lai H. Influences of cholecystokinin and analogues on memory processes. Drug Dev Res 1990; 21:257-276.
5. Lemaire M, Piot O, Roques BP et al. Evidence for an endogenous cholecystokininergic balance in social memory. NeuroReport 1992; 3:929-932.
6. Derrien M, Daugé V, Blommaert A et al. The selective CCK-B agonist, BC 264, impairs socially reinforced memory in the three panel runway test in rats. Behav Brain Res 1994; 65:139-146.

are requested to reveal CCK-8 effects in rats.[12] This has been confirmed by using African green Monkeys and rodents submitted to preinjection stress (reviewed in ref. 77) or novel environment.[78]

All these results suggest that an abnormal increase of the CCKergic activity occurring at the level of mesocorticolimbic DAergic pathway could lead to hypervigilance, pathological anxiety and panic disorders.

## EVIDENCE OF ANTIDEPRESSANT-LIKE EFFECTS INDUCED BY CCK-B ANTAGONISTS: INTERACTION WITH THE ENKEPHALINERGIC SYSTEM

A large body of evidence has now been accumulated supporting physiological interactions between CCK-8 and enkephalins. The distribution of CCK-8 in the brain parallels that of endogenous opioid peptides in numerous brain regions.[79] Moreover, CCK and enkephalins are colocalized in the spinal cord and in several brain structures.[80] It has been suggested that CCK might function as an endogenous opioid antagonist in various nociceptive studies.[7,8,17,81] Furthermore several data showed a tonic or phasic release of endogenous CCK modulating, through CCK-B receptors, the nociceptive threshold by a negative control on μ-opioid receptors at both spinal and supraspinal levels.[9-11,82] Conversely, activation of the endogenous opioidergic system can in turn regulate the extracellular levels of CCK.[83,84]

Interestingly, interactions between endogenous CCK and opioid systems seem also to occur in the control of behavior. Previous experiments have shown that a complete inhibitor of enkephalin-degrading enzymes able to cross the blood-brain barrier, RB 101,[84,85] increased the motor activity, and produced antidepressant-like effects in mice through an interaction of the endogenous enkephalins with the δ opioid receptors.[86] CCK-A receptor activation induced a decrease of the hyperlocomotor activity induced by RB 101 when mice were placed in a familiar environment.[47] However, the most interesting results were obtained with the selective CCK-B agonists, BC 264 and BC 197 in the stressful conditioned suppression of motility test in mice, an animal model used to select antidepressant drugs. These agonists accentuated the suppression of motility in shocked mice and this effect was inhibited by L-365,260. Moreover, the CCK-B antagonist L-365,260 alone decreased motor inhibition in shocked mice. This antidepressant-like effect was suppressed by naltrindole, a selective antagonist for δ opioid receptors suggesting the occurrence of physiological adverse interactions between CCK and opioid systems in behavioral control.[87] Accordingly, the antidepressant-like effect of RB 101 was potentiated by L-365,260 and suppressed by BC 264. As expected, the facilitation induced by L-365,260 on RB 101 responses was blocked by naltrindole[88] (Fig. 8.5).

An antidepressant-type effect was also elicited by L-365,260 in the forced-swim test in mice. This could result from an increase of extracellular dopamine contents since this effect was suppressed both by $D_1$ or $D_2$ selective antagonists and potentiated by co-administration of subthreshold doses of L-365,260 and nomifensine (a blocker of dopamine re-uptake).[89] Taken together these data

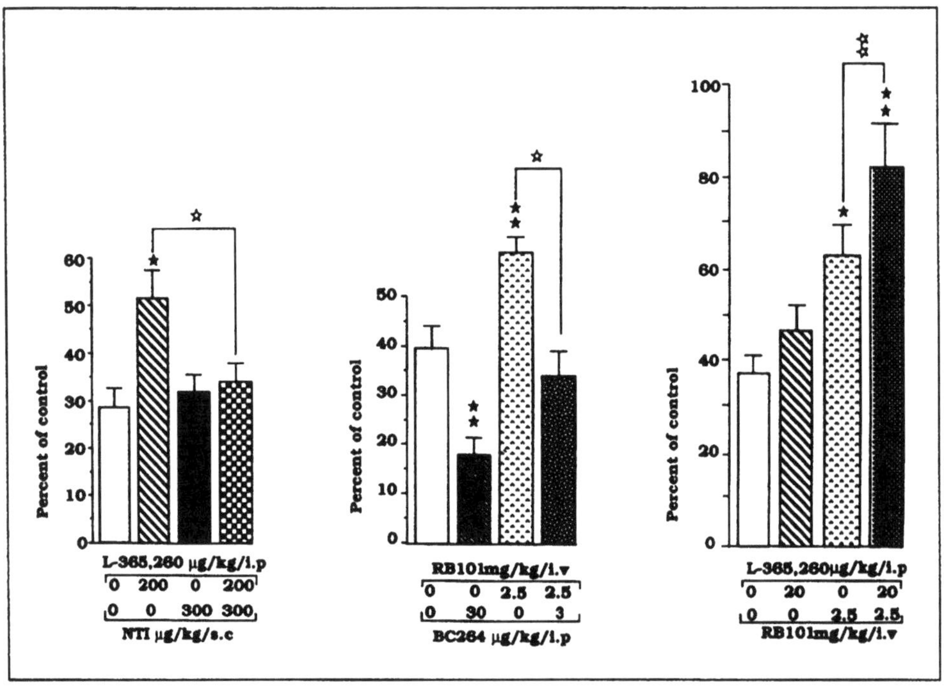


*Fig. 8.5. Antidepressant-like effects of RB 101, a mixed inhibitor of enkephalin degrading enzymes and of L-365,260, a CCK-B antagonist, on the conditioned suppression of motility in shocked mice. The δ opioid antagonist, naltrindole suppressed the antidepressant-like effect of L-365,260. The CCK-B agonist BC 264 increased the conditioned suppression of motility at 30 μg/kg and suppressed the effect of RB 101 at lower dose, 3 μg/kg. The CCK-B antagonist, L-365,260, potentiated the antidepressant-like effect induced by RB 101.*
*L-365,260 or BC 264 was administered ip 30 min before mice was placed in an environment previously paired with inescapable electric footshocks. Naltrindole was administered sc 35 min before the experiment, RB 101 was injected iv 10 min before the test. * $P < 0.05$ ; ** $P < 0.01$, Newman-Keuls test.*

7. Faris PL, McLaughlin CL, Baile CA et al. Morphine analgesia potentiated but tolerance not affected by active immunization against cholecystokinin. Science 1983; 226:1215-1217.
8. Wiertelak EP, Maier SF, Watkins LR. Cholecystokinin antianalgesia: safety cues abolish morphine analgesia. Science 1992; 256:830-833.
9. Noble F, Derrien M, Roques BP. Modulation of opioid analgesia by CCK: evidence of regulatory mechanisms between CCK and enkephalin systems in the supraspinal control of pain. Brit J Pharmacol 1993; 109:1064-1070.
10. Derrien M, Noble F, Maldonado R et al. Cholecystokinin-A but not cholecystokinin-B receptor stimulation induces endogenous opioid dependent antinociceptive effect in the hot plate test in mice. Neurosci Lett 1993; 160:193-196.
11. Maldonado R, Derrien M, Noble F et al. Association of the peptidase inhibitor RB 101 and a CCK-B antagonist strongly enhances antinociceptive responses. NeuroReport1993; 4:947-950.
12. Daugé V, Dor A, Féger J et al. The behavioral effects of CCK-8 injected into the medial nucleus accumbens are dependent on the motivational state of rat. Eur J Pharmacol 1989; 163:25-32.
13. Daugé V, Steimes P, Derrien M et al. CCK-8 effects on motivational, and emotional states of rats involve CCK-A receptors of the postero-median part of the nucleus accumbens. Pharmacol Biochem Behav 1989; 34:157-163.
14. Singh L, Lewis AS, Field MJ et al. Evidence for an involvement of the brain cholecystokinin B receptor in anxiety. Proc Natl Acad Sci USA 1991; 88:1130-1133.
15. Rehfeld JF, Hansen HF. Characterization of preprocholecystokinin products in the porcine cerebral cortex : evidence of different processing pathways. J Biol Chem 1986; 261:5832-5840.
16. Moran TH, Robinson PH, Goldrich MS et al. Two brain cholecystokinin receptors: implication for behavioral actions. Brain Res 1986; 362:175-179.
17. Hill RG, Hughes J, Pittaway KM. Antinociceptive action of cholecystokinin octapeptide ($CCK_8$) and related peptides in rats and mice: effects of naloxone and peptidase inhibitors. Neuropharmacol 1987; 26:289-300.
18. Pélaprat D, Broer Y, Studler JM et al. Autoradiography of CCK receptors in the rat brain using [$^3$H]Boc [$Nle^{28-31}$]$CCK_{27-33}$ and [$^{125}$I]Bolton-Hunter CCK8. Functional significance of sub-regional distributions. Neurochem Int 1987; 10:495-508.
19. Wank SA, Harkins R, Jensen RT et al. Purification, molecular cloning, and functional expression of the cholecystokinin receptor from rat pancreas. Proc Natl Acad Sci USA 1992; 89:3125-3129.
20. Beinborn M, Lee Y-M, McBride EW et al. A single amino acid of the cholecystokinin-B/gastrin receptor determines specificity for nonpeptide antagonists. Nature1993; 362:348-350.

21. Jagerschmidt A, Popovici T, O'Donohue M et al. Identification and characterization of various cholecystokinin B receptor mRNA forms in rat brain tissue and partial determination of the cholecystokinin-B receptor gene structure. J Neurochem 1994; 63:1199-1206.
22. Durieux C, Coppey M, Zajac JM et al. Occurence of two cholecystokinin binding sites in guinea-pig brain cortex. Biochem Biophys Res Commun 1986; 137:1167-1173.
23. Durieux C, Ruiz-Gayo M, Corringer PJ et al. [$^3$H]pBC 264, a suitable 8 probe for studying cholecystokinin-B-receptors: binding characteristics in rodent brains and comparison with [$^3$H]SNF 8702. Mol Pharmacol 1992; 41:1089-1095.
24. Bradwejn J, Koszycki D, Meterissian G. Cholecystokinin-tetrapeptide induces panic attack in patients with panic disorder. Can J Psychiat 1990; 35:83-85.
25. Abelson JL, Nesse RM. Cholecystokinin-4 and panic. Arch Gen Psychiat 1990; 47:395.
26. Bradwejn J, Koszycki D, Couëtoux du Tertre A et al. The panicogenic effects of cholecystokinin-tetrapeptide are antagonized by L-365,260, a central cholecystokinin receptor antagonist in patients with panic disorder. Arch Gen Psychiat 1994; 51:486-493.
27. Bradwejn J, Koszycki D, and Shriqui C. Enhanced sensitivity to cholecystokinin tetrapeptide in panic disorder clinical and behavioral findings. Arch Gen Psychiatry 1991; 48:603-610.
28. Lydiard RB, Ballenger JC, Laraia MT et al. CSF cholecystokinin concentrations in patients with panic disorders and in normal comparison subjects. Am J Psychiatry 1992; 149:691-693.
29. Roques BP, Noble F, Daugé V et al. Neutral endopeptidase 24.11. structure, inhibition, and experimental and clinical pharmacology. Pharmacol Rev 1993; 45:87-146.
30. Fournié-Zaluski MC, Belleney J, Lux B et al. Conformational analysis of neuronal cholecystokinin $CCK_{26-33}$ and related fragments by $^1$H NMR spectroscopy, fluorescence transfer measurements and calculations. Biochem 1986; 25:3778-3787.
31. Charpentier B, Pélaprat D, Durieux C et al. Cyclic cholecystokinin analogues with high selectivity for central receptors. Proc Natl Acad Sci USA 1988; 85:1968-1972.
32. Charpentier B, Durieux C, Pélaprat D et al. Enzyme resistant CCK analogs with high affinities for central receptors. Peptides 1988; 9:835-841.
33. Durieux C, Ruiz-Gayo M, Roques BP. In vivo binding affinities of cholecystokinin agonists and antagonists determined using the selective CCK-B agonist [$^3$H]pBC 264. Eur J Pharmacol 1991; 209:185-193.
34. Durieux C, Ducos B, Blommaert et al. [$^3$H]pBC 264, a highly potent and selective CCK-B agonist for in vitro or in vivo binding studies in various species. DuPont Biotech Update 1994; 9(2-3):52-54.

35. Ruiz-Gayo M, Delay-Goyet P, Durieux C et al. Investigation of opioid and CCK central receptors after peripheral injection of selective and enzyme-resistant peptides. J Control Rel 1990; 13:147-155.
36. Corringer PJ, Weng JH, Ducos B et al. CCK-B agonist or antagonist activities of structurally hindered and peptidase-resistant Boc-CCK4 derivatives. J Med Chem 1993; 36:166-172.
37. Kolodziez SA, Nikiforovich GV, Skeean R et al. Ac-[3- and 4-alkylthioproline[31]]-$CCK_4$ analogs : synthesis and implications for the CCK-B receptor-bound conformation. J Med Chem 1995; 38:137-149.
38. Goudreau N, Weng JH, Roques BP. Conformational analysis of CCK-B agonists using $^1$H NMR and restrained molecular dynamics : comparison of biologically active Boc-Trp-(N-Me)Nle-Asp-Phe-$NH_2$ and inactive Boc-Trp-(N-Me)Phe-Asp-Phe-$NH_2$. Biopolymers 1994; 34:155-169.
39. Rataud J, Darche F, Piot O et al. "Anxiolytic" effect of CCK-antagonists on plus maze behavior in mice. Brain Res 1991; 548:315-317.
40. Hendrie CA, Neill JC. Ethological analysis of the role of CCK in anxiety. In: eds Dourish CT, Cooper SJ, Iversen SD, Iversen LL. Multiple Cholecystokinin Receptors in the CNS. Oxford University Press, New York. 1992: 132-142.
41. Hughes J, Boden P, Costall B et al. Development of a class of selective cholecystokinin type B receptor antagonists having potent anxiolytic activity. Proc Natl Acad Sci USA 1990; 87:6728-6732.
42. Dooley DJ, Klamt I. Differential profile of the CCK-B receptor antagonist CI 988 and diazepam in the four-plate test. Psychopharmacol 1993; 112:452-454.
43. Daugé V, Ladurelle N, Derrien M et al. Potential therapeutic applications of CCK antagonists in anxiety and panic disorders. In: eds Hamon M, Ollat H, Thiebot M-H. Anxiety : Neurobiology, Clinic and Therapeutic Perspectives. Vol 232. Colloque INSERM: John Libbey Eurotext Ltd 1993: 167-176.
44. Blommaert AGS, Weng JH, Dorville A et al. Cholecystokinin peptidomimetics as selective CCK-B antagonists : Design, synthesis and in vitro and in vivo biochemical properties. J Med Chem 1993; 36:2868-2877.
45. Rex A, Barth T, Voigt J-P et al. Effects of cholecystokinin tetrapeptide and sulfated cholecystokinin octapeptide in rat models of anxiety. Neurosci Lett 1994; 172:139-142.
46. Shiosaki K, Lin CW, Kopecka H et al. Boc-$CCK_4$ derivatives containing side-chain ureas as potent and selective CCK-A receptor agonists. J Med Chem 1991; 34:2837-2842.
47. Daugé V, Corringer PJ, Roques BP. CCK-A but not CCK-B agonists suppress the hyperlocomotion induced by endogenous enkephalins, protected from enzymatic degradation by systemic RB 101. Pharmacol Biochem Behav 1995; 50:133-139.

48. Ravard S, Dourish CT. Cholecystokinin and anxiety. Trends Pharmacol Sci 1990; 11:271-273.
49. Sekiguchi R, Moroji T. A comparative study on characterization and distribution of cholecystokinin binding sites among the rat, mouse and guinea-pig brain. Brain Res 1986; 399:271-281.
50. Durieux C, Pélaprat D, Charpentier B et al. Characterization of $[^3H]CCK_4$ binding sites in mouse and rat brain. Neuropeptides 1988; 12:141-148.
51. Knapp RJ, Vaughn LK, Fang SN et al. A new, highly selective CCK-B receptor radioligand ($[^3H][N\text{-methyl-Nle}^{28,31}]CCK_{[26\text{-}33]}$: evidence for CCK-B receptor heterogeneity. J Pharmacol Exp Ther 1990; 255:1278-1283.
52. Derrien M, McCort-Tranchepain I, Ducos B et al. Heterogeneity of CCK-B receptors involved in animal models of anxiety. Pharmacol Biochem Behav 1994; 49:113-141.
53. Palmour RM, Durieux C, Roques BP et al. Anxiogenic effects of CCK agonists in a non-human primate model: central or peripheral ? CCK abstract, 1993 Chatham, Cape Cod, USA.
54. Smith GP, Jerome C, Cushin BJ et al. Abdominal vagotomy blocks the satiety effects of cholecystokinin in rats. Science 1981; 213:1036-1037.
55. Crawley JN. Cholecystokinin-dopamine interactions. Trends Pharmacol Sci 1991; 12:232-236.
56. Smith GP, Gibbs J. Development and proof of the CCK hypothesis of satiety. In: eds Dourish CJ, Cooper SJ, Iversen SD, Iversen LL. Multiple Cholecystokinin Receptors in the CNS. Oxford University Press, 1992: 166-182.
57. Chen DY, Deutsch JA, Gonzalez MF et al. The induction and suppression of c-fos expression in the rat brain by cholecystokinin and its antagonist L-364,718. Neurosci Lett 1993; 149:91-94.
58. Branchereau P, Böhme GA, Champagnat J et al. Cholecystokinin and cholecystokinin receptors in neurons of the brainstem solitary complex of the rat : pharmacological identification. J Pharmacol Exp Ther 1992; 260:1433-1440.
59. Wang Z-J, Rao Z-R, Shi J-W. Tyrosine hydroxylase, neurotensin, or cholecystokinin-containing neurons in the nucleus tractus solitari send projection fibers to the nucleus accumbens in the rat. Brain Res 1992; 578:347-350.
60. Jean A. Le noyau du faisceau solitaire: aspects neuroanatomiques, neurochimiques et fonctionnels. Arch Int Physiol Bioch Biophys 1991; 99:A3-A52.
61. Bradwejn J, de Montigny C. Benzodiazepines antagonize cholecystokinin-induced activation of rat hippocampal neurons. Nature 1984; 312:363-364.
62. Guimaraès FS, Russo AS, De Aguiar JC et al. Anxiogenic-like effect of CCK8 micro-injected into the dorsal periaqueductal grey of rats in the elevated plus maze. In: eds Dourish CT, Cooper SJ, Iversen

SV, Iversen LL. Multiple Cholecystokinin Receptors in the CNS. Oxford University Press, 1992:149-154.
63. Fekete M, Lengyel A, Hegedüs B et al. Further analysis of the effects of cholecystokinin octapeptides on avoidance behaviour in rats. Eur J Pharmacol 1984; 98:79-91.
64. Heidbreder C, Gewiss M, De Mot B et al. Balance of glutamate and dopamine in the nucleus accumbens modulates self-stimulation behavior after injection of cholecystokinin and neurotensin in the rat brain. Peptides 1992; 13:441-449.
65. Daugé V, Böhme GA, Crawley JN et al. Investigation of behavioral and electrophysiological responses induced by selective stimulation of CCK-B receptors by using a new highly potent CCK analog BC 264. Synapse 1990; 6:73-80.
66. Hökfelt T, Rehfeld JF, Skirboll L et al. Evidence for coexistence of dopamine and cholecystokinin in mesolimbic neurons. Nature 1980; 285:476-478.
67. Derrien M, Durieux C, Daugé V et al. Involvement of $D_2$ dopamine receptors in the emotional and motivational responses induced by injection of $CCK_8$ in the posterior part of the nucleus accumbens. Brain Res 1993; 617:181-188.
68. Morino P, Mascagni F, McDonald A et al. Cholecystokinin corticostriatal pathway in the rat : evidence for bilateral origin from medial prefrontal cortical areas. Neuroscience 1994; 59:939-952.
69. Daugé V, Derrien M, Blanchard JC et al. The selective CCK-B agonist, BC 264 injected in the antero-lateral part of the nucleus accumbens, reduces the spontaneous alternation behavior of rats. Neuropharmacol 1992; 31:67-75.
70. Ladurelle N, Keller G, Roques BP et al. Effects of $CCK_8$ and of the CCK-B selective agonist BC 264 on extracellular dopamine content in the anterior and posterior nucleus accumbens : A microdialysis study in freely moving rats. Brain Res 1993; 628:254-262.
71. Marshall FH, Barnes S, Hughes J et al. Cholecystokinin modulates the release of dopamine from the anterior and posterior nucleus accumbens by two different mechanisms. J Neurochem 1991; 56:917-922.
72. Crawley JN, Fiske SM, Durieux C et al. Centrally administered cholecystokinin suppresses feeding through a peripheral type receptor mechanism. J Pharmacol Exp Ther 1991; 257:1076-1080.
73. Hökfelt T, Skirboll L, Everitt BJ et al. Distribution of cholecystokinin-like immunoreactivity in the nervous system : Co-existence with classical neurotransmitters and other neuropeptides. Ann NY Acad Sci 1985; 448:255-274.
74. Zahm DS, Brog JS. On the significance of subterritories in the "accumbens" part of the rat ventral striatum. Neurosci 1992; 50:751-767.

75. Deutch RY, Cameron DS. Pharmacological characterization of dopamine systems in the nucleus accumbens core and shell. Neurosci 1992; 46:49-56.
76. Ladurelle N, Roques BP, Daugé V. The transfer of rats from a familiar to a novel environment prolonges the increase of extracellular dopamine efflux induced by CCK8 in the posterior nucleus accumbens. J Neurosci 1995; 15: 3118-3127.
77. Harro J, Vasar E, Bradwejn J Cholecystokinin in animal and human research on anxiety. Trends Pharmacol Sci 1993; 14:244-249.
78. Lavigne GJ, Millington WR, Mueller GP. The CCK-A and CCK-B receptors antagonists devazepide and L-365,260, enhance morphine antinociception only in non-acclimated rats exposed to a novel environment. Neuropeptides 1992; 21:119-129.
79. Stengaard-Pedersen K, Larsson LI. Localization and opiate receptor binding of enkephalin, CCK, and ACTH/ß-endorphin in the rat central nervous system. Peptides (suppl) 1981; 2:3-19.
80. Gall C, Lauterborn J, Burks D et al. Co-localization of enkephalin and cholecystokinin in discrete areas of rat brain. Brain Res 1987; 403:403-408.
81. Wiesenfeld-Hallin Z, Xu XJ, Hughes J et al. PD-134,308, a selective antagonist of cholecystokinin type B receptor, enhances the analgesic effect of morphine and synergistically interacts with intrathecal galanin to depress spinal nociceptive reflexes. Proc Natl Acad Sci USA 1990; 87:7105-7109.
82. Valverde O, Maldonado R, Fournié-Zaluski MC et al. Cholecystokinin B antagonists strongly potentiate antinociception mediated by endogenous enkephalins. J Pharmacol Exp Ther 1994; 270:77-88.
83. Benoliel JJ, Mauborgne A, Bourgoin S et al. Opioid control of the in vitro release of CCK-like material from the rat substantia nigra. J Neurochem 1992; 58:916-922.
84. Ruiz-Gayo M, Durieux C, Fournié-Zaluski MC et al. Stimulation of δ opioid receptors reduces the in vivo binding of the cholecystokinin CCK-B selective agonist [$^3$H]pBC 264 : evidence for a physiological regulation of CCKergic systems by endogenous enkephalins. J Neurochem 1992; 59:1805-1811.
85. Fournié-Zaluski MC, Coric P, Turcaud S et al. Mixed-inhibitor-prodrug as a new approach towards systemically active inhibitors of enkephalin degrading enzymes. J Med Chem 1992; 35:2474-2481.
86. Baamonde A, Daugé V, Ruiz-Gayo M et al. Antidepressant-type effects of endogenous enkephalins protected by systemic RB 101 are mediated by opioid δ and dopamine $D_1$ receptor stimulation. Eur J Pharmacol 1992; 216:157-166.
87. Derrien M, Durieux C, Roques BP. Antidepressant-like effects of CCK-B antagonists in mice : antagonism by naltrindole. British J Pharmacol 1994; 111:956-960.

88. Smadja C, Maldonado R, Turcaud S et al. Opposite role of CCK-A and CCK-B receptors in the modulation of endogenous enkephalins antidepressant-like effects. Psychopharmacol 1995; in press.
89. Hernando F, Fuentes JA, Roques BP et al. The CCK-B receptor antagonist, L-365,260, elicits antidepressant-type effects in the forced-swim test in mice. Eur J Pharmacol 1994; 261:257-263.

# CHAPTER 9

# Summary and Future Directions

Jacques Bradwejn and Eero Vasar

## SUMMARY

It has been nearly two decades since CCK and gastrin-like peptides were discovered in central and peripheral neurons. These peptides have gained importance in research for several reasons. First, they occur in the mammalian central nervous system at a higher concentration than that of any other known neuropeptide. Second, their distribution is unique, with a highest expression in the cerebral cortex. Third, animal and human studies have suggested a role for these peptides in psychiatric disorders such as anxiety, schizophrenia and eating disorders.

Human and animal studies in support of a relationship between CCK, anxiety and panic and the use of CCK-B agonists as research tools have been summarized in the preceding chapters. Coupland and Nutt have highlighted the requirements for the use of provocation agents in the study of the neurobiology of anxiety and panic attacks. To be useful as provocative agents, compounds need to be safe, mimic spontaneous panic attacks, be specific to anxiety disorders, be reliable and potent and be affected by antipanic agents. CCK-4 meets these requirements and in addition qualifies as a neurotransmitter as discussed in the chapter by Rehfeld and Nielsen. These dual properties give CCK-4 an advantage over endogenous but non-neurotransmitter provocation agents

*Cholecystokinin and Anxiety: from Neuron to Behavior*, edited by Jacques Bradwejn and Eero Vasar. 

such as lactate and $CO_2$ or over pharmacological agents like yohimbine, caffeine and others.

As with other provocative agents, the panicogenic effects of CCK-4 and pentagastrin could be mediated by cognitive factors. However, as summarized in the chapter by Koszycki, studies so far suggest a limited role, if any, for cognitive factors in CCK elicited anxiety/panic. It is nevertheless possible that even though cognitive factors play a minor role on the effects of CCK-4 at doses used in studies so far, its contribution might be apparent when lower or sub-panicogenic doses of CCK-4 are used. This possibility needs to be studied. In any case, even if cognitive factors play a role in CCK-4-induced symptoms, they could be kept constant during neurobiological studies in a laboratory setting.

Harro and colleagues have reviewed animal studies on CCK and anxiety. Manipulations of CCK receptor activity can influence anxiety levels in various species and in very different paradigms, including the elevated plus maze, the light dark exploratory transitions paradigm, the open field, olfactory recognition, social recognition, separation-induced ultrasonic vocalizations, predator-induced freezing, the marmoset threat test, restless behaviors in African green monkeys and the Geller-Seifter punished responding conflict test in rats and squirrel monkeys. In addition, CCK-4 level increases in the brain of rats due to exposure to the odor of a cat. Taken together, animal studies support human findings, indicating that CCK-4 induces anxiety-like behaviors through a CCK-B mechanism.

Harro and colleagues have also proposed that CCK is a normal endogenous anticipatory stress modulator that could be involved in pathological anxiety when the normal regulation of CCKergic mechanisms is altered. In support of this, they point out that neuropeptide transmission is activated by bursting or high-frequency neuronal activity only. Thus, CCK-B receptor activation would elicit pathological anxiety only if the animal is in distress or is temperamentally predisposed to anxiety and that neuropeptide antagonists may be ineffective under normal conditions or in nonpredisposed animals. This postulate would explain the modest behavioral effects of CCK agonists and CCK-B antagonists in animal models as compared to drugs acting on benzodiazepine receptors.

Crawley discussed neurotransmitter interactions with CCK and the possible sites of action on CCK-B agonists in eliciting symptoms of panic attacks. Neurochemical and anatomical hypotheses linking CCK-4 to panic disorder and anxiety are suggested. There is a large body of data concerning the anatomical and functional interaction of CCK with several neurotransmitter systems. The interactions with dopamine, GABA, glutamate, 5-hydroxytryptamine, norepinephrine and opioid peptides seem the most prominent.

A good part of her chapter is devoted to the interaction of CCK with the mesocorticolimbic dopamine system. This interaction is complex and often bidirectional: CCK can potentiate or block the action of dopamine depending on the site of interaction. Although a lot more is known of the interaction between CCK and dopamine than other interactions, its relevance for anxiety and panic expression is not known. A large body of animal literature suggest a role for this interaction in psychosis and substance abuse. Considering the symptomatic overlap between anxiety and these pathologies, it is conceivable that alterations of the balance between the CCK and dopamine systems in specific brain sites ie mesocortical/mesolimbic regions unselectivley contribute to different syndromic expressions.

Several lines of evidence suggest functional interactions between CCK and GABA, and the direction of this interaction seems to be mostly antagonistic across paradigms. CCK also influences the activity of glutamate but this interplay is more controversial as CCK seems to potentiate as well as to block the effects of glutamate. Two other monoamine neurotransmitters, 5-hydroxytryptamine and norepinephrine, interact with the CCK system. A reciprocal relationship between 5-hydroxytryptamine and CCK seems to be at play. CCK seems to potentiate the anxiety-related actions of 5-hydroxytryptamine and CCK antagonists may block the effects of 5-hydroxytryptamine in dorsal raphe neurons. Similarly, norepinephrine could play a part in the inhibition of CCK agonist-induced anxiety. Finally, CCK and opioid peptides antagonize each other in the physiology of pain sensitivity and CRF might mediate the anxiogenic action of CCK.

In summary, the interactions described above could contribute to endogenous anomalies of CCK functions and to the expression of anxiety or panic induced by CCK-B agonists.

The chapter by Vaccarino is devoted to the role of CCK in reward and in anxiety due to withdrawal of abused drugs. He implies that CCK may be involved in mechanisms underlying anxiety processes associated with withdrawal states. Also discussed is evidence in favor of a contribution of CCK to mechanisms known to mediate the incentive value of drugs of abuse. The involvement of CCK in both anxiety and reward processes raises the possibility that a component of the anxiogenic properties of CCK reflects CCK actions in the nucleus accumbens and influences the neuronal processes related to reward. It is interesting to speculate that CCK, through action on CCK-B receptors, could cause anxiety and alterations of reward activity in part by its ability to depress dopaminergic activity. In light of the opposing role of CCK-A and CCK-B receptors, CCK action on CCK-A receptors may produce an effect which may or may not be associated with anxiety. It may be that CCK-A receptor activations contribute to the acquisition of psychostimulant reward by enhancing dopaminergic function under conditions of drug-induced dopamine activation. It would be particularly interesting to examine the extent to which CCK-A receptors play a role in the sensitization to psychostimulant drugs, a process thought to be critical in the development of drug reward.

V. Daugé and B.P. Roques described in depth the interaction between enkephalins and CCK. These two peptidergic systems seem to be activated in several stressful and conflicting situations in humans, monkeys and rodents. The release of CCK related peptides could lead to various behavioral responses such as nociception and anxiety through interaction with CCK-B receptors. While the physiological determinants of their release remain speculative, studies performed with inhibitors of enkephalin catabolism, and CCK and opioid antagonists show that CCK and enkephalins participate in opposite manner in the elaboration of adaptive behaviors. The enkephalinergic system appears to produce opposite effects through activation of $\mu$ and $\delta$ opioid receptors respectively. The potential area of overlap between fear, anxiety, panic, emotionality and pain, which has been vastly understudied, could be approached by using CCK has an investigative tool in anxiety disorders and in acute and chronic pain states.

## FUTURE DIRECTIONS

Compelling questions about the role of the CCK system in anxiety and panic remain and several investigative strategies can be envisioned to answer them. Although there seem to exists concordance between animal and human research, as justly pointed out by Harro and by Crawley, caution must be taken against over-interpretation of animal data. The relevance of various animal models of anxiety to human research on fear, anxiety and panic in humans is not known. While there are very good animal models for anxiety-related behaviors which predict anxiolytic drug response in humans, there are presently no clear rodent models specific for panic attacks and panic disorders. As human expression of anxiety and panic attacks is largely subjective, validation of animal models might only come from pharmacological effects of anxiolytic and antipanic drugs. Additional studies of the behavioral effects of CCK agonists in primates might provide models which are equivalent to human model of panic induction.

In order to sustain clinical relevance, a fruitful strategy could be to use the CCK provocation model in humans to set directions for concerted investigations using both human and animal models. Even if the phenomenological concordance between behaviors elicited by CCK in humans or animals remains unclear, both approaches can feed each other. For instance, animal research can bring information about the involvement of neurotransmitter systems in CCK-induced behaviors when pharmacological dissection would not be possible in humans because of the lack of availability of molecules safe for use.

## THE FOLLOWING POINTS SEEM IMPORTANT TO INVESTIGATE

### Selectivity of action of CCK-B agonists in anxiety disorders

It is still unknown whether CCK plays a role in panic disorder alone or whether it is involved in the pathogenesis of other anxiety syndromes. Establishing whether CCK-4 is specific to panic disorder could potentially improve the diagnosis of anxiety disorders and perhaps prevent their occurrence. As previously mentioned, panic disorder and normal control subjects display marked differences in their sensitivity to the panicogenic effects of CCK-4. Between 75-90% of patients panic with a 25 μg challenge dose of CCK-4 compared to less than 20% of healthy volunteers. This represents a difference in panic rate ranging from 55-70%. Although the effects of CCK-4 in patients suffering from other psychiatric disorders have yet to be determined, it is possible that lower doses of CCK-4 will differentiate panic disorder patients from those with other pathological states. Additional provocation studies could help determine whether CCK-4 shows selectivity of action in panic disorder and whether it could be used as a diagnostic tool.

**Aberrations of the CCK system in anxiety disorders**

It is possible that the enhanced sensitivity to CCK-B agonists of patients with panic disorder results from an abnormal ratio of CCK-4 and CCK-8S due to an anomaly of function or expression of metabolic enzymes. Clarification of the biogenesis and metabolism of various CCK fragments could provide tools for neurobiological studies of anxiety and panic. Comparisons of the biogenesis and metabolism processes of CCK in healthy volunteers and anxiety patients could help uncover anomalies at any level(s) of these processes. Along this line of reasoning, association between endogenous levels of gastrin and anxiety-linked personality traits has been suggested. It will be important to assay blood levels of CCK-4, versus blood levels of gastrin, under baseline conditions in panic patients. Moreover, critical, but difficult, experiments would be to assay blood levels of CCK-4 and gastrin during spontaneous panic attacks, in panic patients. Such assays could determine whether levels of CCK-A and CCK-B receptor agonists are temporally correlated with symptoms in panic patients.

Genes encoding the processing of CCK, and expression of CCK-A and CCK-B receptors have been delineated. Accordingly, studies in molecular genetics could determine whether genetic expression of CCK fragments or of CCK receptors could be altered in anxiety patients and whether this could help explain the genetic transmission of these illnesses. There may be alterations in the nucleotide sequence in the CCK-A and/or CCK-B for genes that confer enhanced sensitivity of CCK-B receptors and vulnerability to panic disorder. Studies of polymorphism of CCK-A and CCK-B receptor genes could be performed to reveal possible alterations in patients suffering from panic disorder. Also, the close proximity of dopamine and CCK receptor genes on human chromosomes gives pertinence to molecular genetic studies of these two systems in anxiety disorders, psychosis and addiction.

In animals, knock-out strategies of site directed injections of antisense deoxyoligonucleotides against the formation of CCK in the brain, and the expression of CCK-A and CCK-B receptors seem worthwhile. Complementing these studies could be investigations of gene over-expression for the production of CCK-related peptides and receptors in any given region of brain.

The observation that panic disorder patients are hypersensitive to the panic-promoting effects of CCK-B agonist and that they exhibit decreased CSF and lymphocyte concentrations of CCK-8S (a mixed CCK-A and CCK-B agonist) may suggest that panic disorder is associated with an imbalance between CCK-A and CCK-B systems. As discussed earlier, electrophysiological and behavioral studies in animals suggest that selective CCK-A and CCK-B receptor antagonism can produce opposite behavioral effects and sites such as the brainstem nucleus tractus solitarius (NTS), which receives projections from primary sensory afferents of the vagus nerve, appear to be crucial sites for the action of CCK-A and CCK-B agonists. The enhanced sensitivity of panic disorder patients to CCK-4 could be explained by a hyperactivity of the CCK-B receptor system or by a underactivity of the CCK-A receptor system in the brainstem. This hypothesis could be tested by comparing the relative effects of CCK-A and CCK-B antagonists on CCK-4 panic attacks in humans.

**Effects of CCK-B antagonists on spontaneous panic attacks or noninduced anxiety**

A major support for the role of CCK in panic or anxiety disorders would be provided by a demonstration of therapeutic efficacy of a CCK-B antagonist in panic disorder or generalized anxiety disorder. So far, adequate tools to tests the CCK hypothesis of anxiety with clinical trials have not been available as the only two compounds adequate for human use, L-365,260 and CI-988 show unfavorable pharmacokinetic properties. Fortunately, compounds with superior pharmacokinetic profiles exist and are intended for clinical trials. However, it should be emphasized again that, even if results of clinical trials are negative, the CCK-4 challenge paradigm could nevertheless remain a practical research tool that can enhance our knowledge of the neurobiological mechanisms subserving panic attacks and lead to the development of novel antipanic drugs.

**Elucidation of the neuroanatomy subserving anxiety and panic attacks**

It is unknown whether CCK molecules can cross the blood-brain barrier but studies so far seem to indicate that the effects of

CCK-4 could be the result of its action across the blood-brain barrier. The central/peripheral site of action for CCK-4 is an important question. However, considering that the anatomical sites of anomalies or abnormal functions responsible for panic attacks are unknown, identification of any site of action of CCK-4 and pentagastrin could by itself bring more invaluable information about the pathophysiology of panic than answering the question of a central verses peripheral effect. Imaging techniques with radioactive markers will probably play an important part in future investigative strategies. Concerted human and animal research strategies could attempt to pinpoint sites of action of CCK-4 enumerated by Crawley in her chapter.

**The neurotransmitter network of anxiety or panic in humans**

CCK-B agonists could be used to map out neurotransmitter networks leading to the phenomenon of anxiety and panic attacks. Strategies of pharmacological dissection of the panicogenic action of these agonists could be used in humans with agonists/antagonists of postulated CCK-4 target systems. In this fashion, the role and relative contribution of these neurotransmitter systems to the phenomenon of anxiety and panic could be determined. New tools might thus be developed. In addition, studies of endocrine correlates of the effects of CCK-B agonists in human and animals could help our understanding of the role of hormone release in anxiety, panic, stress and arousal.

Site-specific studies of CCK and neurotransmitter interactions could be performed. For instance, CCK-4 administered into the ventral tegmentum and/or prefrontal cortex would be predicted to induce panic-like and/or anxiety-like behaviors in rodent and primate models. While a large number of behavioral studies have been performed to date with CCK-4 and CCK-B antagonists, these animal experiments employed systemic or intraventricular routes of administration. Direct microinjection of CCK-4 into regions of the mesocortical dopamine pathway, in sensitive animal models relevant to human panic responses, would be a straightforward method to evaluate the role of CCK and its interactions with neurotransmitter systems in the production of anxiety or panic.

Noradrenergic projections of the locus ceruleus have been implicated in the detection of novel stimuli, in vigilance and anxiety-related behaviors and in stress responses. An interesting question is whether CCK-4 activates locus ceruleus neurons, with a resulting contribution to the panic response. Experiments on the effects of CCK agonists on the firing rate of locus ceruleus neurons would be important to address the hypothesis of CCK-nonadrenergic interactions relevant to CCK-4-induced panic.

It is interesting to speculate that CCK-4 administration could activate a nociceptive pathway, e.g., by blocking enkephalinergic systems in peripheral organs, spinal cord and/or brain sites. Interoceptive stimuli generated by a decrease in opiate tone could then produce further internal cues, which are interpreted by the patient in a manner which leads to panic symptoms. This hypothesis could be directly tested by pretreatment with an opiate agonist, to increase opiate tone, which would be predicted to prevent CCK-4-induced panic. Alternatively, it would be interesting to test whether naloxone, a standard opiate receptor antagonist, produces panic-like symptoms in panic patients and in normal controls.

The hypothesis of CCK-glutamate interactions as mechanisms for CCK-4-induced panic asserts that CCK-4 activates a glutamatergic pathway which mediates panic symptoms. This hypothesis could be tested directly in animal models in which CCK-4 produced anxiety-related behaviors and cardiovascular effects. If selective glutamate receptor subtype antagonists blocked CCK-4-induced behavioral and physiological actions, then the anatomical site(s) mediating glutamate/CCK-4 interactions could be analyzed through site-specific microinjections.

Last but not least is the interaction between CCK and GABA. These neurotransmitters are co-locolized in several key structures in the brain which control emotions and cognitive processing. Brain site directed injections of CCK and GABA selective drugs, and applications of recent developments in molecular biology will give new insights to the role of interaction between CCK and GABA in the regulation of fear and anxiety. The effect of CCK antagonists in benzodiazepine and alcohol withdrawal in humans might be also provide interesting insights.

# INDEX

*Page numbers in italics denote figures (f) or tables (t).*

## V



## W



## Y

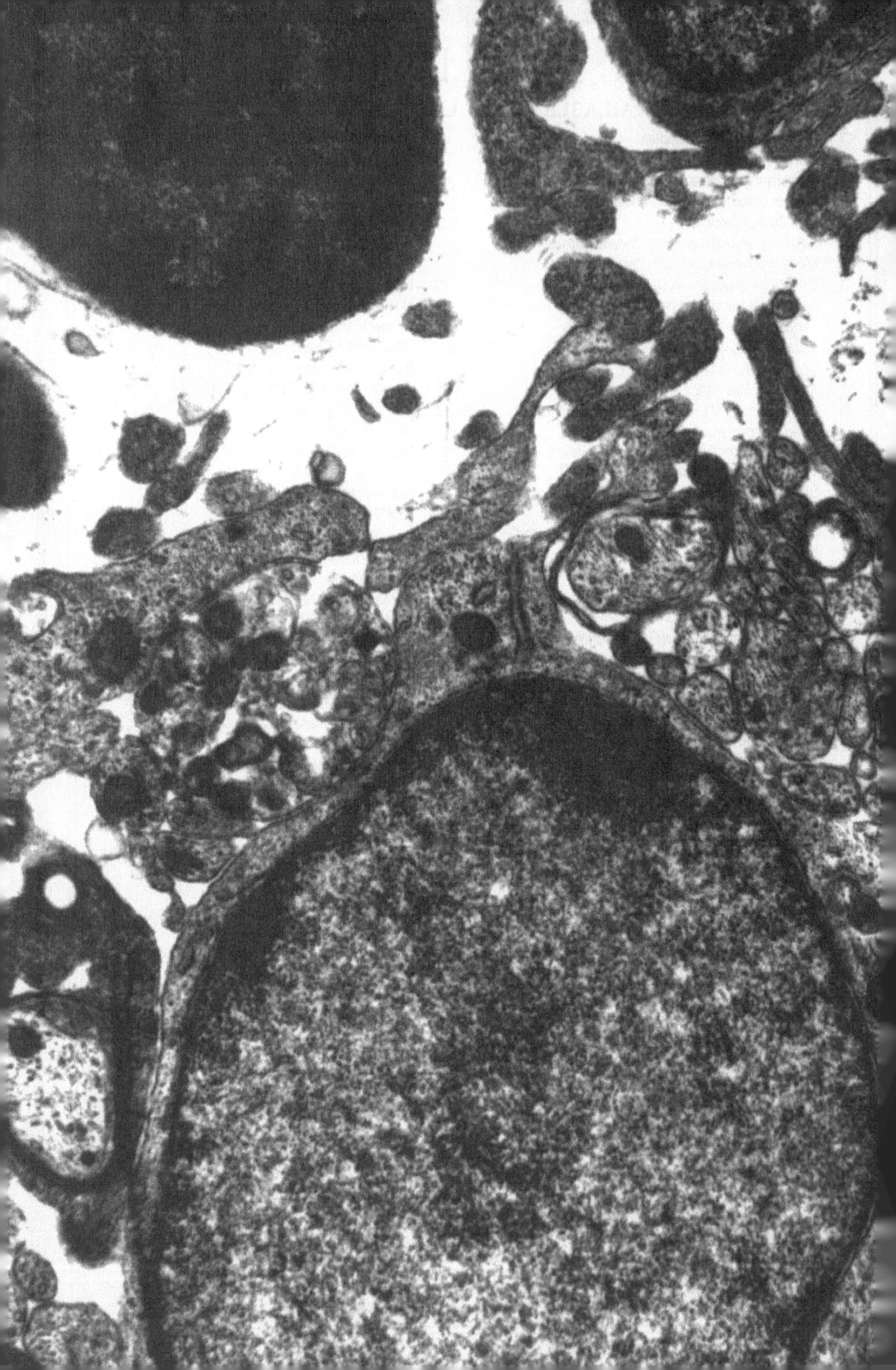

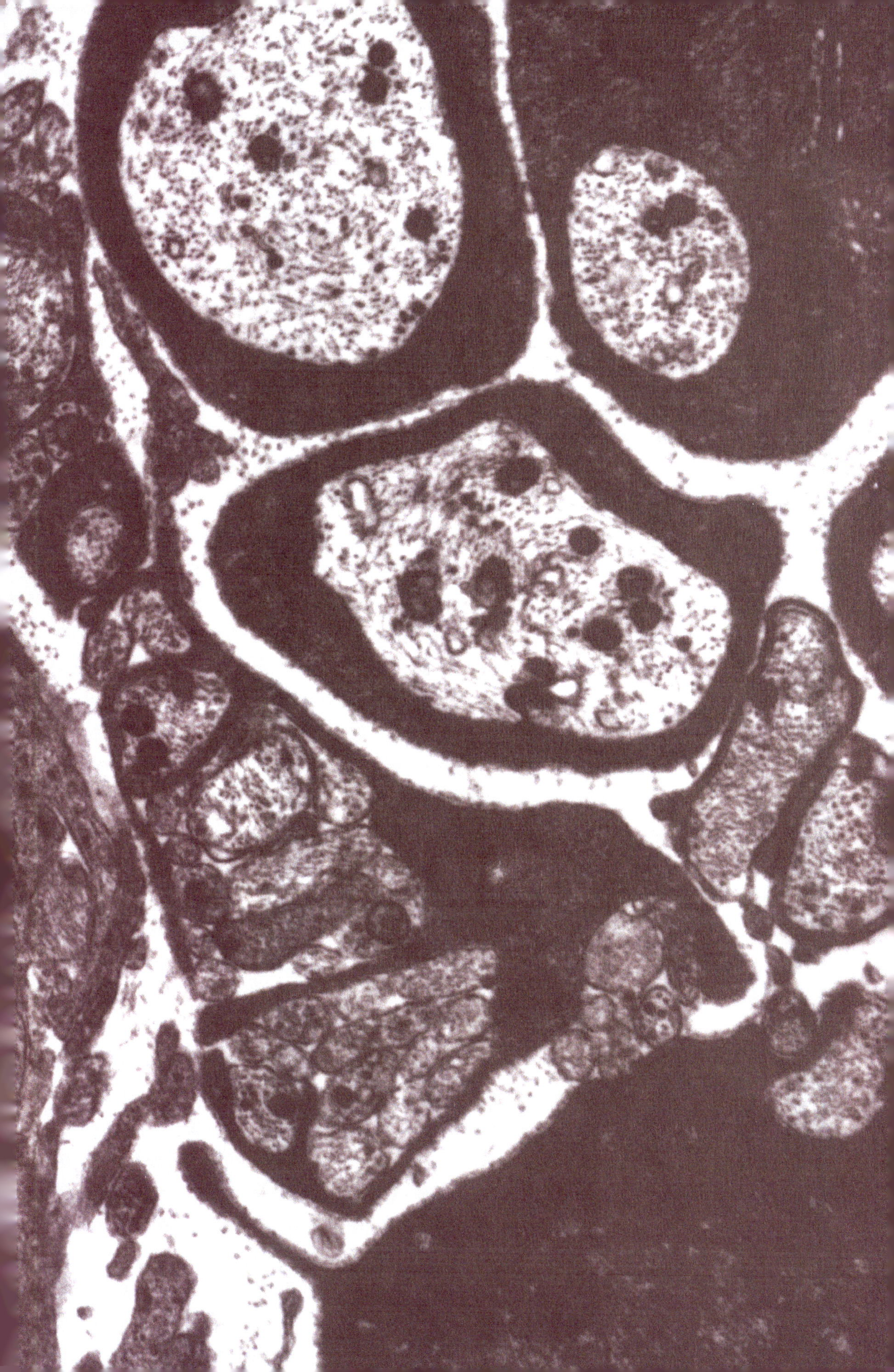

MIX
Papier aus verantwortungsvollen Quellen
Paper from responsible sources
FSC® C105338

If you have any concerns about our products,
you can contact us on
**ProductSafety@springernature.com**

In case Publisher is established outside the EU,
the EU authorized representative is:
**Springer Nature Customer Service Center GmbH**
**Europaplatz 3, 69115 Heidelberg, Germany**

Printed by Libri Plureos GmbH
in Hamburg, Germany